Elegant Technology

Economic Prosperity from an Environmental Blueprint

Jonathan A. Larson

The Riverdale Company, Publishers

Published in 1992 in the United States by The Riverdale Company, Publishers, P.O. Box 2679, Westwood, Massachusetts 02090

ISBN: 0-913-215-58-9
LOC: 92 064031

Printed and bound in the United States of America

10 9 8 7 6 5 4 3 2 1

TABLE OF CONTENTS

PART ONE:
The People of Industrialism

CHAPTER ONE
THE HUNTER AND THE FARMER 1

CHAPTER TWO
MARXISM: THE FAILED PROMISE 31

CHAPTER THREE
ARE PRODUCERS A CLASS? 39

CHAPTER FOUR
THE THEORY OF THE INDUSTRIAL CLASS
WHEN CREATIVITY BECOMES IMPORTANT! 51

PART TWO:
The Economics of Industrialism

CHAPTER FIVE
THE ECONOMICS OF PREDATION 77

CHAPTER SIX
MONEY! 97

CHAPTER SEVEN:
FORDISM 121

CHAPTER EIGHT
TOOLS! 127

CHAPTER NINE
THE GREAT RIVERBOAT GAMBLE 145

PART THREE:
The Industrial Environmental Solution

CHAPTER TEN
DO PRODUCERS HAVE A PLAN? 165

CHAPTER ELEVEN
TECHNOLOGICAL ELEGANCE 191

CHAPTER TWELVE
BUILDING A CONSENSUS FOR ACTION 219

CHAPTER THIRTEEN
THE HURDLES TO
INDUSTRIAL ENVIRONMENTALISM 239

CHAPTER FOURTEEN
THE POPULIST PERSPECTIVE 259

PREFACE

SOME COURAGE IS REQUIRED to link the notions of technology and industrialization to the hope for environmental renewal. Industry has been portrayed as the bad environmental actor for so long that the concept of industrial environmentalism seems a hopeless contradiction of terms.

Industrialization has been blamed for many great evils—environmental destruction is only the last great crime. The holocaust of Fascism, the nuclear incineration of Hiroshima and Nagasaki, the horrors of colonialism, and the insanities of Stalinist Marxism were all made possible or infinitely worse by industrialization. At times one wonders if industrialization is not the genuine original sin. The more interesting questions concern temptation—What about industrialization is so attractive that it has been relentlessly pursued whenever humans have understood its possibilities? What about industrialization is worth fighting wars over?

Before industrialization, it took months of high adventure to cross the American continent by foot, horseback, and canoe. Now the same crossing takes six hours and the biggest hazards are bad airline food and the fear of lost luggage. No one wants to sew or saw without electricity once a Skilsaw or sewing machine has been tried. And certainly no one wants to return to the days of preindustrial medicine when most women eventually died giving birth and 60 to 90 percent of all children died before the age of five.

Going backward toward a preindustrial state is made impossible because preindustrial skills have been forgotten. Few farmers still know how to grow crops using horses and the few remaining horsedrawn implements are in museums. Even the Amish, who have maintained a preindustrial lifestyle for religious reasons, find it very difficult to remain outside of industrialization. Even with a powerful cultural tradition that teaches new generations the old skills, these skills are so many and difficult that often Amish youth leave the sect, not so much because the seductions of the modern world are so enticing, but because they simply cannot master preindustrial skills. Anyone who has seen an Amish barn-raising knows that these skills are not trivial. Amish barns are built with factory-produced nails, so even these people are not as purely preindustrial as they would wish themselves to be. Imagine a typical urbanite, who has difficulty lighting his hot-water heater, becoming five times more self-sufficient than the Amish, and the scope of the difficulty of returning to our preindustrial past is illuminated. We cannot go back because we simply no longer know the way.

The seductions of industrialization are many for ultimately, industrialization is, and has always been, about possibilities. For most, industrialization is about their very survival. The question must be asked, How did the drive for survival turn so mutant that it now threatens to destroy the biosphere?

Industrialization is about thermodynamics, and like the fire that drives it, it is a cursed blessing. It can and has been used for good and evil.

Industrialization is about tools. The motor of industrial change is the relentless human striving for perfection. Perfect goods can only be built with perfect tools and the perfect industrial organization of work. The crimes attributed to industrialization are a foul perversion of a sublime human characteristic.

Because industrialization is about fire and tools, the crime of ecocide is a sin of industrialism itself. It is the producing class version of hubris. Industrialization is about human creativity—the assumption that the gifts of nature needed creative transformation. When measured against the creativity of nature, the creativity of human beings is still quite primitive.

Humans may protest that their creativity is not so primitive. A compact disc player, Mercedes Benz, or bioengineered medicine must be at least as sophisticated as an amoeba. Point taken. Yet human creations at their most sophisticated do not reproduce by themselves. Until a television set can mate and reproduce itself, humans are perpetually responsible for their creations. With industrialization the rule is—in for a nickel, in for a buck. There is no going back with the advanced products of industrialization—the nuclear technologies of the 1950s, alone, will be a burden to humanity for 10,000 years. There is no back-to-nature solution for radioactive waste. There is only one solution for industrial problems—get more sophisticated industry, very quickly.

The other superior quality of an amoeba is connectedness. The amoeba lives and dies as a link to all other parts of nature. Since an automobile cannot eat worn out television sets and give birth to a new car, humans must manage the links between their creations as well. This is the environmental imperative to industrialization—the industrial loop must be closed.

Industrial environmentalism is the social imperative to industrialization—your creativity must mimic natural creativity. If that means your technology must become more elegant and your production more sophisticated, then that is a part of the imperative as well.

This is what *Elegant Technology; economic prosperity from an environmental blueprint* is all about.

Part One: The People of Industrialism, examines the history and social development of the peoples who produced the industrial state. In spite of the fact that the United States is an industrialized country, the people and problems of industrialization remain almost invisible. The struggle over the definition of industrial organization has always had profound social implications. This struggle has defined most of the twentieth century—its wars and ideological combat. The end of the Cold War is the perfect time to reexamine the core questions of industrialization, stripped of its irrelevant distractions, now that 75 years of missing the point are behind us.

Part Two: The Economics of Industrialism, reviews the cultural philosophies of the people who invented twentieth century industrialism and contrasts them to the neoclassical economic theologies taught in American universities. After two decades of world-wide economic and social fundamentalism, there exists a crying need to recognize that the imported views of David Ricardo, Karl Marx, and Adam Smith are not very relevant.

If American problems are to be solved, it is time to teach the economics of the people who invented the United States. The political-economic theories of Benjamin Franklin, Thomas Paine, Thomas Jefferson, Frank Lloyd Wright, Thorstein Veblen, Henry Ford, the Populists, Robert M. La Follette and the progressive Republicans are certainly relevant if the industrial state must be reinvented. Combined, the political-economic theories of these great thinkers could be called THE ECONOMICS OF INDUSTRIAL DESIGN. It is the economic thinking necessary to produce an industrial environmental society.

This thinking, oddly enough, is already manifest in the economies of our former enemies—Japan and Germany. They have already made the economic commitment to an industrial-environmental future using lessons we taught them. Japan's postwar central bank was organized by a Detroit banker. It is not surprising that Japan's banks have successfully funded the North American invasion of Japan's automobiles.

American industrial redesign needs new economic understanding as much as better environmental thinking. Fortunately, some of the industrial-economic philosophies of the thinkers at the birth of the current manifestation of industrialization shed considerable light on the path to industrial environmentalism.

Part Three: The Industrial Environmental Solution, looks at the existing examples of elegant technologies and speculates on the social and economic needs for a graceful transition to an industrial form that conforms to the imperatives of the natural order, and seeks to answer the questions: Can the damaged environment be rescued from total ruin by a more sophisticated industrial philosophy? Can industrial societies be governed by philosophies that are preindustrial? When is industrial planning appropriate and when is it not? Does industrial environmentalism have core principles and if so, what are they? Are successful industrial societies by necessity, by definition, high-wage, full-employment societies? What are the implications for the twenty-first century of an industrial-environmental strategy?

Thank You!

Writing a book is a particularly solitary pursuit. Yet human support is welcome and at times, critical. I would especially like to thank the following for their assistance:

Robbin Zoupas who provided sustenance for this project before I was even sure what I was doing or why I was doing it;

Frank Ryan, my tool guru, whose regular reports from the front lines of industrial production proved invaluable;

Neil Heidemann, the consummate craftsman, whose love of beauty inspires awe. Whenever I needed a perfect living example of a Michelangelo, Neil fit the bill—he even paints beautiful ceilings;

Bill Haight, who always supplied inspiration to continue; and

Jonathan Carlson, the film documentarian, who taught me that it is not necessary to know for what you are searching—keep looking anyway. As a compulsive book collector, Jon bought dozens of the writings of American Midwestern progressive thinkers from the period of 1870 to 1940. His collection of these works may now be one of the best and they were absolutely critical to much of what is in my book.

Elegant Technology; economic prosperity from an environmental blueprint has already been published in Scandinavia. It first saw the light of day in 1989 as *Tuottajat ja Saalistajat: Johdatus ekoteolliseen ratkaisuun* (Producers and Predators: An introduction to industrial-environmentalism). How this book came to be published first in Finnish is a long and interesting tale—too long to include here. The people who made it happen were kind, warm, and generous. A special thank you to :

Pekka Sulkunen, a brilliant and perceptive social writer, professor, thinker, and friend, who found my book a publisher and insisted that it be translated. He wrote the Finnish foreword where he introduced me as the logical heir to the teachings of C. Wright Mills. Such a compliment inspired redoubled efforts;

Heikki Salojärvi, a witty and conscientious translator, whose editing gave me interesting insights into the concerns of Scandinavian policy intellectuals. His translation is reputed to be excellent—he may have made my book better than it was originally;

Matti Virtanen, the man with a plan, the writer, editor, and publisher at *Hanki ja Jää,* who took a big chance on an unknown American writer simply because I had a program too. Scandinavian societies owe their successes to dedicated servants to the public good such as Virtanen; and,

Marjatta Jaatinen, a complex, gentle, and caring woman, with an artist's soul and the fire to match, who patiently taught me the important subtleties of Nordic culture and promoted my book with a believer's intensity. Her ideas, determination, and beliefs inspired and shaped the long and difficult days of the rewrite.

Thank you. All of you. You helped me survive and gave meaning to my life. This book was not possible without your help.

Part One:

The People of Industrialism

CHAPTER ONE

THE HUNTER AND THE FARMER

IN THE BEGINNING, there was agriculture. Before agriculture, humans were barely different from the other primates. Agriculture would change everything because agriculture grows more than plants and animals: it grows civilizations. Before agriculture, human groups consisted of nothing more than wandering clans in search of food. With agriculture, humans could predict when and from where their food would come. Having solved this essential problem, humans would go on to build cities and libraries and governments.

Unlike many scientific discoveries that would follow, the person or persons who discovered agriculture will never be known. Hundreds of theories, many plausible, have been advanced to explain the coming of agriculture. It could have been discovered because of a burial ritual where grain placed with a corpse began to sprout. The accidental scattering of gathered grains by the wind or a broken container could have led to farming. Undigested grains passing through animals will grow new plants—a primitive and accidental form of agriculture. No matter the scenario, a critical event took place. Someone was able to distill the essence of an accidental happening so that it could be recreated on purpose. Crossing the bridge from noticing and defining phenomena to causing phenomena is the quintessential act of human genius. Whoever noticed that seeds grow plants had to convince others to bury carefully gathered food in the hope that more food would come. That person was not only a first-rate scientist, but something of a promoter as well.

In many ways, agriculture is still an act of genius. Many derogatory terms have surfaced through the years to label those engaged in farming. "Peasant" and "serf" are not words of flattery, but slander will never obscure the fact that farmers were some of the first scientists and to this day, many of the greatest scientists, engineers, and inventors have their roots in agriculture. It is possible that genius of this form is an inherited trait and the descendents of the inventors of agriculture are simply redeploying their skills in design shops and laboratories; or it is possible that the agricultural environment is a particularly fine instructor as to the laws of nature. In either case, agriculture was caused by and has caused a new strain of humanity that is in many respects different from the hunting and gathering clans from which farmers sprang.

Agriculture, like any other invention, was clearly not for everyone. The skills necessary to be a great hunter such as daring, aggressiveness, speed, good eyesight, and cunning are not very well suited for farming. In fact, a person can be a very good farmer by being cautious, defensive, plodding, nearsighted, and honest. It probably helps.

No matter the advantages conferred by this marvelous new invention called agriculture; some people still preferred to hunt. They were good at it and they enjoyed it. It only stands to reason that even with the invention of agriculture, many people were more suited for hunting. Any suitable hunting characteristics that can be genetically transmitted had long since become a dominant strain through the processes of natural selection. If a man had the great depth perception and the hand-eye coordination necessary to throw a spear accurately, he was more likely to survive to pass those traits along to his offspring. Some nomadic clans never did acquire agriculture and stuck to hunting, which is what they knew best.

For those nomadic tribes that never figured out agriculture, the agricultural plots they happened upon merely became concentrated sources of food that they took for themselves. The farmers, faced with a loss of their crops, entered into a fateful agreement. They would pay the good hunters in their own clans to defend against the raiders. In return for forty percent of the crop, for example, the hunter-defenders would protect the farmers from those who would take everything. Sixty percent of the crop is better than nothing.

How long this arrangement lasted, no one knows; probably not very long. It could not have taken the hunter-defenders very long to realize that those who farmed had neither the time, skills, nor inclination to defend themselves. The hunter-defenders took over. They merely shifted their predatory activities from harvesting the bounty of nature to seizing the production *and the producers* they were supposed to protect. The geniuses who had invented agriculture now found themselves a form of property of the hunters.

Although they have a great deal in common, the ability to invent farming and the ability to farm are not the same thing. Historically, the producers began to stratify along the lines of creativity.

Valor and bravery are the signal virtues of the hunters while the ability to transform nature creatively is the signal virtue of the producers. If the highest members of the hunters are called heroes, the highest members of the producers are called geniuses. (There have been military men called geniuses but in fact, "military genius" has always been an appropriated expression that is, in fact, an oxymoron.) Although those producers with lesser abilities and skills wound up with the jobs of shoveling out the animal shelters, those with great creative skills eventually left agriculture altogether. This did not mean that they had entered the ruling class. Rather they became the highly skilled artisans and artists who lived better than those left behind on the farm.

At the top of the creative heap were the weapons-makers. Weapons have existed far longer than agriculture. The few remaining nonagricultural tribes have very simple weapons; but "simple" is the key word. The process of creative stratification of the agricultural producing-groups means that specialists were now making weapons where before everyone made his own. Since predators were in charge of matters, those with the greatest creative skills usually made weapons. When the first full-time weapons maker appeared is open to question. Probably the Egyptians had such persons. The Greeks may have had them. The long-boat makers of the Vikings could be considered full-time weapons makers.

It took until the nineteenthth century before enterprises appeared with no other function but arms manufacture. Naturally, the weapons of the twentieth century are the most sophisticated, destructive, and dangerous. While the specialists of the agro-artisan groups have learned to build atomic weapons, the hunters have learned nothing. Most would be hard put to fabricate a bow, a simple skill their forebears had. The combination of a bow-and-arrow mentality and nuclear weapons is history's most frightening. What is even more peculiar is that most significant improvements in weapons were made by a producer, such as Alfred Nobel, who deluded himself into thinking that using his invention in war would be so horrifying that none would dare.

Because the hunters have been perpetually in power since the dawn of civilization, most of recorded history is about the activities of the hunters. History books are filled with wars and battles and intrigues. It is odd that the group of people who made most of the history possible, the agriculture-artisans, is almost never mentioned. We know the names of the great Viking sea-captains. We do not know the name of their boat builders. We do not know who invented the stirrup or gunpowder, or paper, or other history-changing invention. The agriculture-artisan class (after this called the producer class) is not given a fair hearing in the history books, but this does not mean that the history of the producer class is unimportant. History has always been written by the flatterers of the powerful and it should not be surprising that the history of the weaker producer class should be ignored, even if that history is vastly more interesting and important to understanding modern problems.

The producer class has made occasional attempts to affect its own lives in ways other than changing what they could make and grow. It was inevitable that producers try other methods because no matter what the producer classes accomplished in terms of technical and productivity gains, the results were taken by the predator class. What the producer classes needed was a social and political philosophy around which to organize. The producer classes got its first such philosophy with the arrival of Jesus of Nazareth.

That Jesus of Nazareth was a producer-class teacher cannot be doubted. Raised in the home of a carpenter, he went on to surround himself with followers from the exploited groups in his society. His teachings are laced with producer-class examples. In the story of "The Good Shepherd," Jesus of Nazareth tells the story of how a hired shepherd abandons the flock of sheep when the wolf attacks. The good shepherd is the farmer who has an interest in his sheep because he is the owner. The good shepherd does what the hired hand will not do and spends the night gathering the flock. (John 10:11-15) [1]

As a producer-class illustration, the story of "The Good Shepherd" is without peer. Since the time of Jesus of Nazareth, many ways have been tried to organize agriculture. No method of organization has ever come *close* to yielding the productivity of the owner-operator method. When production decisions in agriculture are made by someone other than the person doing the farming itself, whether by absentee owners or by bureaucratic commissars, agricultural production is depressed and misery increases.

Other reasons stand out to explain why Jesus of Nazareth was the first great producer-class giant. Foremost are his teachings of peace. The philosophy of the producer class is that when conditions of peace prevail, production will rise and there will be plenty for everyone. The disruptions of warfare are so great that anything gained will not offset the losses of production. The desire for peace is not only the central fixture of the producer mentality, it is a distinctly antihunter position. The more Jesus of Nazareth talked of peace, the more nervous the men in power became, and eventually they did what people in power do with irritants: they had him killed.

The peaceful producers of the world now had a hero-martyr who had said that the peacemakers would be called "Sons of God." (Matthew 5:9) Jesus of Nazareth had not been gone very long before the decision was reached to "beef up" the philosophy a bit. The disciple Paul—a disciple has best been defined as one who gets an "A" in the course and misses the whole point[2]—began to talk of Christians as soldiers. (Ephesians 6:11-17) Women, who had formed an important element of Jesus of Nazareth's following, were now told to keep silent in church. (1 Corinthians 14: 34-35) Early Christian experiments with production communes were tried and then abandoned. That the philosophies of Jesus of Nazareth, who came from the home of a builder, and those of Paul, who was a member of the religious establishment, should differ substantially is not surprising but inevitable given the basic antipathy of those backgrounds.

By the third century A.D., Christianity had lost much of its steam as a revolutionary movement. Christianity was adding a bureaucracy that was modeled after the Roman system. As time progressed, Christianity became what religion had always been, an excuse for those in power to remain in power. The meekness and passivity of Christian thought no longer threatened the hunters who cynically exploited the people whose imaginations had been illuminated by the thoughts of peace. Christianity became just another excuse for warfare and predatory behavior.

Christianity remained in this dismal state until the Protestant Reformation. Martin Luther fired the producer-class imagination by saying that every person has the right and duty to communicate personally with God—thereby bypassing the oppressive and predatory religious bureaucracy. Luther is also famous for pointing out that a person's worth is not contingent on the job held or the position in society, but by how well one does one's assigned job. Such a notion put the hard-working and productive peasant higher up the social ladder than the lazy aristocrat. Luther lived his life as if he believed such a notion, but when the peasants took him too seriously and revolted, Luther found himself siding with the princes.

Because modern Lutherans are such practical, orderly, and nondemonstrative citizens, it is easy to forget the radical nature of Luther's teachings. This is a mistake. The Lutheran countries of Scandinavia are as socially advanced as any on earth—demonstrating past social victories. Currently, Lutheran clergymen and theologians were central to *Neues Forum*—the movement in old East Germany (D.D.R.) that was as responsible as any for the elimination of the Berlin Wall. *Neues Forum* was as German Lutheran as Polish *Solidarnösc* was and is Catholic. The nationalist separatist leaders in Latvia and Estonia are often Lutheran and look to their Scandinavian cousins as models in many matters. [3]

Menno Simon, the Anabaptist leader who founded the Mennonites, never compromised producer principles. His followers became committed pacifists and organized their religious beliefs around nonpredatory production techniques and notions of shared provision. Prohibitions on lawsuits and holding government office showed the breadth of Menno Simon's understanding that predatory practices extend far beyond the organized armed robbery that is warfare. Violent men have gone out of their way to prey on this group.

Producer notions were at the heart of the American Revolution. The intellectual progenitors of the Declaration of Independence, Benjamin Franklin and Thomas Jefferson, had impressive producer credentials before they become involved with political movements. Both were inventors and Franklin was a first-rate scientist. Jefferson's notions as to how agriculture was to be organized were much in line with Jesus of Nazareth's ideas of owner-operators. Jefferson referred to his agricultural producers as yeomen-farmers who were the backbone of democracy.

Like the producer movements that preceded it, the American Revolution soon fell into the hands of the hunter class. What Paul did for Christianity, Alexander Hamilton accomplished for the American Revolution. To understand the mindset of Hamilton and his ilk, remember that he and others attempted to get Prince Henry of Prussia to become king of the United States in 1786. [4] By the time Shay's Rebellion had been put down, the producer classes in America found that the great notions of the American Revolution had been abandoned and that rule from Washington was not substantially different than it would have been from London. There was one difference, however. Jefferson and Franklin were, and still are, heroes of the Revolution and absolutely above reproach. Their insights were so clear and their understanding so profound that two hundred years later, their ultimate invention, the United States, is still a marvel, though their design was only partially followed.

The producers needed more than religious and political thinkers, important as they were, to overturn their subservient position in relation to the hunter class. They got such a man in Abraham Darby, the first great English steelmaker. Darby was a pacifist Quaker and hence found himself outside the bounds of English society. He was denied the "privilege" of a university education. He learned his trade in the brass industry because the steelmakers of the day were attached to the arms business. A freedom from preconceived notions allowed Darby to innovate extensively in the production and marketing of steel. [5]

That Darby was a Quaker is no accident. The essential characteristic of the Protestant faiths is religious anarchy as opposed to the authoritarian hierarchy of the Roman church. The Quakers, with their notions of following the light within, which is something akin to a well-developed conscience, are in a class by themselves. They are the logical outcome of the Protestant Reformation. More simply, they may be the first Christians fully to understand Christ.

Because producers avoid fighting, they are often viewed as cowards. This is not so. The bravery of a predator is demonstrated by physically exploring new turf. The producer demonstrates bravery by intellectually exploring new territory. The terror of thinking new thoughts must be very real for the few who ever attempt it. Trusting the light within provides comfort when opening the intellectual doors that lead to intellectual and creative freedom. After all, we refer to a creative insight as a spark or flash and ideas are represented by a light bulb in the mind. It is the same light.

For moral reasons, the Quakers refrained from engaging in the weapons business. They got out of the big-church business.

The two routes to the top of the producer heap were blocked for Quakers of ability. Combine the innovative thinking allowed by the trust in the light within with restricted outlets for such thinking and the industrial revolution was practically inevitable—where else was the talent to go?

It would be difficult to overstate the importance of the Quakers to history. It is generally conceded that they were responsible for the industrial revolution. As David B. Davis points out in his *Slavery and Human Progress*, the Quakers were the progenitors of the abolitionist movement and kept it alive during all its many setbacks. In the nearly 2000 years since Jesus of Nazareth, there have been only two social revolutions that have ever really improved the human condition: the industrial revolution and the abolition of slavery. The Quakers were largely responsible for both.

Because the Quakers were outside the English establishment, the industry they represent has always suffered from a discriminatory bias in British society. Industry was just something a gentleman did not do. This anti-industrial bias has destroyed the British industrial underpinnings in the twentieth century and threatens to do the same in the United States where all things English are considered wonderful, no matter how useless or irrelevant.

What makes Darby, and the other early steelmakers, such as John Wilkinson, so interesting is their response to the cash shortage that threatened to stifle growth. Their workers had to be paid in cash and not in kind. The noblemen of England were not sure about this new notion called industry. When they failed to mint enough coinage to support the new ironworks, the ironmongers minted their own coins with their unnoble likenesses stamped on them. No wonder it was called the industrial *revolution.* [6]

The industrial revolution threatened the hunter power establishment by threatening the very definitions of wealth. No longer was wealth defined as land with serfs attached to it. Wealth became defined as those heretofore worthless resources that were now necessary to feed the maw of emerging industries. Wealth was now counted as money and the tools of production—collectively known as capital.

The hunter class is nothing if not cunning. The new industrial enterprises were soon owned by the predator class and became as exploitive as the old agricultural arrangements had ever been. Instead of a bright new day, the industrial revolution brought the producer class more ghastly working conditions.

The rise of industrialized production spawned social critics and philosophers to explain why increased production caused such widespread poverty and misery. The producers, of course, knew the reason: predators were taking too much and this had the side effect of destroying the ability to make societies rich.

The most influential of these social critics were the economists—those who Robert Heilbroner calls "The Worldly Philosophers" in his highly literate book of the same name. Heilbroner's list of economic philosophers whose ideas changed the world extends from Adam Smith to John Maynard Keynes. That all but one of Heilbroner's "philosophers" wrote in England is not surprising since the industrial revolution began there. What is ominous about the list is that all but one have no experience, inclination, or sympathy for the problems of production—not even Marx.

If Heilbroner's list is definitive, it would appear that economics as a philosophy is usually an attempt by someone to describe industrialization to predators using the language, perspectives, and superstitions of the hunter. Producer-economist is an oxymoron except for one stellar exception: a first generation Viking-American named Thorstein Bunde Veblen who had deep roots in agriculture and an unexcelled ability to discern the difference between the important and absurd manifestations of the emerging industrial state. [7]

THE PRODUCER'S WORLD ACCORDING TO VEBLEN

THORSTEIN VEBLEN is in many ways the definitive industrial-class philosopher-economist. He was uniquely qualified by time and location for this role because he was able to observe at first hand all the steps necessary to transform a society from a preindustrial to an industrial state.

Born in 1857, his childhood was spent on a farm that was a model of early industrial agriculture. In addition to being recognized as the most innovative farmer in the county, his father was also a master carpenter who had learned his trade as a youth in Norway. As a result, young Thorstein grew up with a profound understanding of the rituals and techniques of the Scandinavian handicraft traditions. While the American industrial revolution blossomed around him as a university student at Yale, Veblen became interested in the differences between the reality of industrial production and the predatory commercial Calvinist theologies of school.

Veblen's writings focus on the difference between the values of those who produced the industrial revolution and those who stole it from them. He taught at the University of Chicago, the school bought with John D. Rockefeller's ill-gotten gains, which gave him unique access to the latest currents in the justifications for the practices of industrial predatory fraud.

Veblenian scholarship often overlooks an interesting facet of his understanding of the producer-predator conflict. His father, though far from devout, was culturally a Lutheran who defined himself as *not a Calvinist.* [8] In what may be Veblen's most prescient work, *Imperial Germany and the Industrial Revolution* (1915), he describes the cultural differences between England and Germany in ways that Calvinists and Lutherans alike would understand. [9] By learning Lutheran teaching at home while attending schools—Carleton and Yale—that were little more than thinly disguised Puritan seminaries, Veblen had absorbed both traditions and understood their power and hazards. For him, World War I was little more than an unfortunate civil war between the two dominant wings of the Protestant Reformation—a conflict made inevitable because of their differing approaches to the industrial revolution.

In *Imperial Germany*, Veblen's astute analysis predicts most of the important events of the twentieth century. At the risk of oversimplification, *Imperial Germany* argues that because Germany imported the industrial revolution from England, she was culturally unfit to comprehend her new-found industrial muscle. The industrial revolution, though invented by social outsiders, grew from enlightened English tolerance. The enlightenment missed Germany and after the Revolution of 1848, many of her most skilled workers fled to England where they learned the new industrial forms. Most would return during the reign of Bismarck and bring their newly acquired skills and ideas but by then, Germany was forced to play industrial catch-up.

Veblen noted that although Germany had been culturally unfit to *begin* the industrial revolution, she was better fit than England to improve on and sustain its potential. In Bismarck's Germany, the industrial revolution was supported by the establishment. Because of official policy and a more cohesive society, industrialization in its productive forms would never suffer the outcast status it did in England. Nevertheless, while Bismarck's Germany would embrace, enhance, and develop the productivity of industrialism, the nation did not accept industrialism's most important cultural premises: intellectual freedom and pacifism.

By 1915, Germany had passed England in any reasonable comparison of industrial might, but industrialism would confuse German culture. Feudal Prussian, Lutheran, Junkerian, Bismarckian subservience, blind loyalty, and patriotism mixed with industrialism caused, according to Veblen, the outbreak of World War I. German unification under Bismarck had brought together the arrogant militarism of Prussia and the might of Krupp's Ruhr.

Veblen's analysis postulated that industrial values grow from industrial practices. Accordingly, England was the home to the industrial revolution and was more peace-loving than Germany where the new industrial values had yet to take root. Even so, the industrial west of Germany was already less inclined toward war with England than the preindustrial Prussian east. He predicted that until the industrial values superceded Prussian patriotism completely, Germany would cause her neighbors great trouble.

The internal cultural dilemmas of Germany's industrial revolution spilled over into the two most deadly wars of history. During the second conflagration, German industrialists acted, or were complicit in, a systematic attempt to coerce or murder everyone not contributing to or defending the new industrial state. By 1939, Germany's army had in many ways become a servant of industrialism. The German armies were destroyed in an attempt to seize industrial raw materials and exterminate the enemies, high and low, of the industrial order.

With militarism destroyed in battle and the philosophic remnants of predation eliminated, Germany would emerge from World War II as a model of enlightened industrial behavior. Today, in the country that produced the feared *Wehrmacht*, 50,000 young men a year register as pacifists and German civilian industrial output is the envy of most of the world. *Die Grünen,* the German Green Party, has succeeded in imposing environmental constraints on traditional industrial practices. The resulting industrial environmentalism represents the purest of Nordic industrial values. Veblen would have approved of modern, technologically sophisticated, antimilitarist Germany.

Veblen might be stunned today by the extent of modern English deindustrialization but his analysis predicted that as well. He fixed 1913 as the year Germany passed England in industrial power and assumed this repositioning of the industrial order was irreversible because England had effectively stopped competing.

The twentieth century has demonstrated that conflict between producers and predators can be a bloody ordeal. The outcome, moreover, is still far from clear. When John Kenneth Galbraith wrote the *New Industrial State* in 1967, it appeared as though industrialism had become dominant internationally. Then, under the banner of Thatcherism, preindustrial, imperial, predatory, Victorian capitalism made a curtain call in the 1980s. England would export only ideas this time, and they would be the antithesis of the industrial revolution. Having deindustrialized, England would encourage others to follow her lead. The English-speaking world and many other nations did just that.

While never properly defined, Thatcherism became a blueprint for industrial and economic decline. It was claimed that high interest rates do not matter, industry and industrial workers are no longer important, low wages are good, industrial planning is impossible or evil, and that production is a problem reserved for the lower classes and those who lose wars. "Honor" is a reason for warfare, but industrial assets will be sold to anyone with anything that resembles money. Financial transactions are to be watched with profound interest; the health of the real economy is ignored. Bond traders have wealth and status while engineers and scientists stand in unemployment lines.[10] Most of all, nothing can ever be thought wrong in the economy as long as the shopping malls are full of goods. Consumption will be king: production is deposed.

The Thatcherite industrial counter-revolution of the 1980s rallied about the flag of "free trade." Because everyone wants both freedom and trade, this idea steamrolled all objections. It should have been remembered that the free trade flag was flown in the Opium Wars of the nineteenth century. The English colonial drug dealers, in their attempt to use opium as social control, encountered Chinese objections. Opium may have made easier British rule but drug addiction was ruining China—especially her cities. Opium was being produced in British India and shipped to China. China tried to close her borders to opium.

These were not properly called wars—it was a one-sided affair in which millions of Chinese were massacred or impoverished. When the extent of the carnage—over what was essentially a drug deal gone bad—became known the rest of the world, the keepers of British manners and self-esteem were forced to produce a high-sounding excuse for destruction and mass murder. For the history books, the Opium Wars were fought for the lofty ideals of free trade: the right of commerce to operate in a world without borders.

Free trade originally meant in practice that England got to set the world's economic rules to her advantage. Under the slogan of free trade, England seized port facilities and took over customs stations. England would build infrastructure improvements to facilitate her plunder—such as docks and railroads from harbors to mines, bill the colony, and send in the military to collect the debts.

Free trade was the catch-all justification for colonial excess. Free trade came to mean that colonies had no say in the operation of their economies. Americans can easily understand a colony's objection to such a situation. It was over the issues of economic autonomy that the American Revolution was fought. We wanted to run our economic lives.

Because free trade is a philosophy of theft, it is a recipe for decline. While England grew fat on her plundered wealth, and extracted the last farthing from her hellish Midland factories, contempt and neglect of productive enterprise caused her to rest on her industrial laurels. All the important groundwork in organic chemistry was done in England. All the important chemical companies in Europe are German. Pure science may be an honorable occupation for an Englishman of leisure, but turning science into industry is frightfully expensive and time-consuming, and the sort of work fit for—harumph—a German, for example. It may take time, but eventually hard working and clever people live better than thieves. Flying the banner of free trade, England lost her empire and is now an irrelevant industrial actor with a lower standard of living than almost any country of western Europe.

Free trade's Thatcherite manifestation was almost as ugly as the first version. In the 1980s, free trade came to mean that no area had a right to defend its standard of living, that workers must compete only with the lowest-paid labor on the globe, that geography and climate are of no importance in how a society chooses to organize agriculture, and that citizens do not have the right to defend ecological areas. Under the banner of free trade, developing nations were ordered to destroy their social fabric in the name of global finance. The bankers were quite clear, "Our greed is insatiable so we have raised the price of our product—money. You will not spend for those frivolous items in your budget, like typhus and cholera vaccines. You must send the money to us instead."

The idea that trade must have no rules is absurd. Trade rules are often flawed and political boundaries are seldom the same as rational trading areas, but there are limits to how far and for what reason molecules in any form should be shipped. Veblen complained loudly about the tariffs and other trade restrictions of his day—mostly because they hurt the farmers of the Internal Empire, America's Midwest. He thought it wicked that Germany had designed her trade rules so that she could be completely self-sufficient in a war. According to Veblen, mutual need for trade between nations produced peace so it must be encouraged rather than hindered for the gain of *predation*. He recognized the power of trade legislation for good *and* ill—the rules must be mutually beneficial.

Valid trade barriers include the protection of agriculture, intellectual property, vital industrial infrastructure, worker's rights, and the environment. Every nation or region must insist on these barriers. Trade rules are important—too important to be left to those who believe that because it is difficult to write good trade rules, no rules are better. Like the ideas of a flat earth, the gold standard, and communism, the monster of free trade has done far more damage than good and should be buried in an unmarked grave.

THE DEINDUSTRIALIZATION OF AMERICA

THE AMERICAN anti- and deindustrial frenzy of the Reagan era swept away all opposition in its path. Under the catch-all banner of free trade, Republicans and Democrats, Liberals and Conservatives, consumer radicals, feminists, environmentalists, bankers, and antimilitarists united in one voice to cheer the demise of the industrial state and its traditions. Industrialism was blamed for Fascism, wars, pollution, over-population, and according to Robert Bly, the death of maleness itself.

American "free traders" organized the export of the heart and brains of American industrialism: post-industrial ideology became a preindustrial strategy. Industrial dominance, the prize of World War II, became in 1980s America a worthless commodity to be disposed of as scrap. The industrial sum of three generations of intense struggle became ignoble.

The ideological commitment to the glories of preindustrial free trade was astonishingly pure and took upon itself absurd manifestations. One example was most telling. In 1985, in a scene from out of Mitchener's *Hawaii*, James Fallows, the liberal, articulate, Harvard-educated editor of *Atlantic Magazine,* took it upon himself to relocate in Japan so as to more effectively spread the gospel of free trade.

Fallows got it backward. Missionaries, secular Calvinist or otherwise, are supposed to journey from the "mother country" to the colony. His hopeless task was to convince the Japanese that a free trade economic philosophy (which had plunged the American economy from unchallenged industrial colossus in 1945 to the status of world's biggest debtor in 40 years) was superior to their system. Fallows returned home to rethink his religion. Like many missionaries that have gone before, Fallows "went native" and has become the foremost American scholar on Japanese culture, but while he was away, the religion spread to the farthest corners of Anglo-American influence.

The scope of anti-industrialism even reached Minnesota, the childhood home of Veblen and the modern home of innovative and influential Minnesota Mining & Manufacturing (3M). Minnesota was probably more immune than most places to the absurd contradictions of deindustrial strategies. Yet even there, in the name of "economic development" during the mad reign of Reagan, the state issued taxpayer-backed *Industrial* Revenue Bonds for the purpose of building a horse-racing track. Horse-racing, the sport of kings, the highest flowering of preindustrial amusement, had become an "industry." It is to be wished that this was the most important example of the deindustrialization of America! It was not.

As in most of America, most Minnesota Industrial Revenue Bonds were issued to encourage real estate development. This would have hardly surprised Veblen. In *Imperial Germany,* he comments that government involvement in real estate speculation was a huge impediment to American industrial development. He complains that the building of the railroads was so motivated by real estate considerations, the trains and tracks themselves were near-worthless junk. For Veblen, infatuation with real estate development is the distinguishing characteristic of those seeking preindustrial economic power.

By 1945, the Germans and Japanese had learned the hard lesson that industrial dominance has nothing to do with territory, militarism, or imperialism. So while they spent the 1980s investing in industrial dominance, the deindustrializing Americans reverted to real estate speculation. Japan got the commercial technologies of the twenty-first century for her investment. America got empty office buildings for hers.

Germany and Japan would mostly ignore the tide of deindustrialization. Losing a war completely had purified their industrial philosophies by destroying the hunter classes. Rebuilding a destroyed industrial society would enhance the prestige of the builders. Cultural self-assurance would allow them to stand up to preindustrial English cultural imperialism.

As the 1980s progressed, the world's deindustrializing economies cracked under the strains of predation. In the 1990s, culturally-dominated producer industrialism, embodied in the cultures of modern Germany and Japan, stands triumphant over Victorian English capitalism.

In the Anglo-American world, there was no ideological defense of industrialism. Industrialism is simply not the sort of ideology that has many defenders. Heilbroner may have been correct in his analysis that Veblen was the only producer philosopher-economist in the English Language. In England, producers have always been considered a class not given to philosophizing.

American producers do not face the class and cultural barriers of their English peers. The lack of modern industrial philosophers to mount a defense of American industrialism is better explained by a frontier form of pragmatism which states, "Real producers must produce something tangible—only dilettantes have time to write or philosophize about production." Veblen published a huge body of complex and difficult writing in his life, yet the farmers who knew him considered him quite lazy.

Validating Veblen's foresight, twentieth century industrial-class geniuses have created huge, wealth-generating enterprises that have benefitted virtually every human who has come in contact with them—the owners, the towns they were in, the employees, the merchants, and the customers. Names like Thomas Edison and Henry Ford, Edwin Land of Polaroid, William Norris of Control Data, Donald Petersen of Ford Motor, and Stephen Jobs of Apple Computer are perfect examples. In Pehr Gyllenhämmär of Volvo and Akio Morita of Sony, we have in the 1990s, genuine international industrial-class superstars.[11]

In a sense, these people could be called "producer-economists." None is, of course. One of the industrial revolution's salient features is the division of labor. People who use new thinking to organize production, implement it, and thereby change the world's economies are not called economists. That designation goes to those who write about the changes in industrialization after the fact.

There are some American economists who represent the necessary thinking of production in spite of the tradition of economics as the philosophy of the predators. These mavericks include John Kenneth Galbraith—the prolific neo-Veblenian Harvard economist who began his life on a farm, and Lester Thurow who labors for Massachusetts Institute of Technology (M.I.T.)—arguably the intellectual epicenter of American industrialism. Producer writers on economic themes, such as Buckminster Fuller, Amory Lovins, and Edwards Deming, are not considered economists.[12] This is especially odd in the case of Deming because his economic theories are practically worshipped in Japan—the world's most successful industrial society today.

Veblen pointed out in 1915 that, as in England, American industrialization is the product of social outsiders. Henry Ford, the dominant industrialist of the early twentieth century, was Irish and as far outside the established order as was possible—even in his home town of Detroit. Non-English immigrant groups have contributed significantly to American industrial progress. During this century, Germans directed the space program and invented much of the computer hardware, Italian and Jewish immigrants unleashed nuclear power, and in the 1980s, industrial entrepreneurs from around the world have redefined what remains of American industrialism. Unfortunately, the American social, economic, and political establishment has never seen industrialization as anything but wealth to be plundered. Industrial might meets social power on a regular basis and because England was used as a model in the 1980s, industry lost influence. As deindustrialization creates the social catastrophe of economic collapse, producer philosophy will regain its lost legitimacy and the producer-economists will be heard once again.

The philosophies and practices of industrialism have never become totally dominant in the United States. Cultural confusion about industrialism in America means that she resembles in many ways, 1915 Germany: still looking for a war to fight, still hoping the military will organize consumption of industrial output. While America dithers about industrialism, the nation loses ground to those cultures where the issues have been settled. America is being forced to make a decision about the commitment to industrialism because no decision is worse than a bad one.

The predators choose to ignore or discredit industrialization because they have been having a lovely twentieth century. Two world wars conducted at insane levels of brutality plus countless lesser atrocities, cataclysmic economic upheavals, and an international system of codified usury that makes the exploitation of nineteenth century British colonialism look like benevolence, all combine to let the world know that the hunter-predators are still out there. Unfortunately, predatory tendencies that arose through the millions of years when those were just the traits to have, will not disappear by themselves in a few thousand years, certainly not in a few hundred.

Japan's industrial history is similar to Germany's. Like Germany, Japan imported industrial ideas and tools. Deming did not become Japan's industrial-producer guru in a vacuum. World War II had three effects on the Japanese hunter-predator classes. [13] First, they lost a war and were killed or discredited. Second, General MacArthur saw to it that the industrialists were granted the right to reorganize. He insisted that they do away with their petty bickering and get on with a serious program of standardization. Third, their industry had to involve itself strictly in nonmilitary enterprises.

The peace terms forced upon Japan created a genuine producer class state. Japan accomplished the job in only 40 years. More interestingly, Japan's huge economic size and phenomenal technical accomplishments have finally attracted the attention of the great predator powers—the United States and what is left of the crumbling Soviet states.

Of the teachings of Jesus of Nazareth, there has been one ridiculed more than any other, "Blessed are the meek, for they shall inherit the earth." (Matthew 5:5) There is a passage in the musical *Camelot* where the line has been changed to ". . .it's not the earth the meek inherit, it's the dirt." John Paul Getty, the oil billionaire, is supposed to have said, "The meek may inherit the earth—but not the mineral rights."

While no one who has competed recently with the Japanese in business would call them meek, the fact remains that compared to the belligerent, ruthless, arrogant fighters the Japanese were only a few decades ago, they are now polite, decorous pacifists by comparison. That is close enough to meek for this rough outline of history. Not only is Japanese influence greater now than at the zenith of the militaristic Japanese-East Asian Co-prosperity Sphere, the Japanese own the most modern production facilities world-wide.
14

In September of 1985, when the United States became a debtor nation, much financial power shifted from New York to Tokyo. Some might argue that because Japan's financial markets are not up to world leadership, Tokyo is still not the financial center of the planet. Industrial leadership will solve this minor objection—sooner rather than later. This is not the same as inheriting the earth, maybe, but it's very, very close.

The emergence of a producer superpower like Japan is sufficient reason to treat the producer class with the importance that it deserves. This is much more than a matter of fairness. The producer class, although not warlike, has some problems of its own—even when it solves the social and economic riddles of industrialization. The ultimate dilemma of the industrial revolution is environmental destruction with its mounds of waste and constant resource depletion.

It should not be surprising that since the twentieth century saw the first real rise in power of the producer class, many major, intractable problems of the late twentieth century industrial economies are producer problems. The problems of the producer class can only be solved by the producer class. The problems of toxic waste cannot be solved by war or revolution. To get to the heart of industrial problems, it becomes more important than ever that producers, their ideas, and their history not be ignored.

Because there is no escaping the environmental dilemmas posed by industrialization, America must *choose* an industrial environmental strategy. So far, the United States has aped English deindustrialization. Industrialization has been declared the environmental problem to be addressed with legislation to regulate production, funds for "clean-up" squandered on legal hand-wringing, and industrial liquidation through financial fraud—all examples of *non*-decisons. The invisible hand became unconscious.

England survived deindustrialization quite gracefully—America will not. Too many Americans define themselves by their work to be declared useless burdens who should wander out onto the ice, like an old Eskimo, and die. England has had a few riots. The United States risks civil war if deindustrialization continues—everyone is *already* armed and violently angry. The suicides, assassinations, and violent demonstrations associated with the American agricultural collapse of the 1980s should warn of the impending social chaos—now that the economic collapse has migrated to the cities, industries, and financial institutions.

The choice of deindustrialization in America is odd. In spite of pockets of Anglophilia, Americans do not like the English or their thinking very much. Whenever the United States acts like old imperial England—such as when overthrowing governments in Chile, Iran, Guatemala, and Panama, official Washington is forced to act in secret for fear of public outrage. American imperial pretensions have been so controversial within the United States that thousands owe their employment to *denying* them. Americans dislike English industrial output so much that they refused to buy the Rover Sterling automobile—even though it was nothing more than a Honda assembled in England. Imitating England is strange for a country that fought a Revolution to free itself from English rule.

In contrast, German and Japanese industrialization is widely admired in the United States—their products are status symbols. Americans believe, moreover, that Japanese-German industrialization is *American* industrialization. "We taught them everything they know," is a sentiment widely expressed and to some extent, accurate. Americans who understand production find Japanese and German industrial practices stunning.

The American *public* does not need to be convinced that the German-Japanese industrial model is superior, but rather the people who run the banks, teach the children, operate the bureaucracies, and pass the laws. It is this lagging socio-economic indicator that must be convinced to look at another alternative. It is they who must be convinced that industrial-economic might has little to do with armies. It is they who must be convinced that industrialization has many social organizations and that Marxism was the enemy and alternative to but one of them.

The time is right to look more closely at German and Japanese industrial ideas. The end of the cold war means that these pacifist countries have become prototypes of civilian production. If, as it appears, Germany has *already* invented an environmentally-correct industrial strategy, the last valid impediment to an American conversion to a more elegant industrial order has been eliminated. American industrialization will never be identical to a German or Japanese version—the cultures are too different, but, fortunately, a new American industrial order could be superior. Germany and Japan both taught that the student can best the teacher. It is to be hoped that American industrial environmentalism becomes the world's finest—for American environmental problems are the planet's most serious.

NOTES ON CHAPTER ONE

[1] The New Testament of the Bible. (King James Version)

There are several ideas lifted from the Bible in this book. This has been done for valid reasons. It is impossible to ignore the impact of Christian thought on industrialization. The vast majority of the pieces of the industrial structure are the products of Christian cultures.

In the future, as Japan becomes even more important to the industrial landscape, the teachings of Buddhism, Confucianism, or Shintoism may guide their

NOTES ON CHAPTER ONE

development and will become a necessary subject for discussion—I hope someone qualified is already working on such a book.

Japanese social structures affect their industrial organizations—but even here, many important ideas appear borrowed. In many ways, the Japanese social organizations of industrialization are a better representation of Protestant Germany than the Lexus 400 is of the Mercedes Benz. There is an obvious link between what is made and how it is made that spills over into cultural areas such as religion. The more interesting question concerns causation: Do core values produce the culture which can build a Mercedes, or does the ability to imagine a Mercedes create the core values?

The included reference markings apply to all derivations of the King James Bible. The choice of the King James Bible was quite deliberate. There are many Americans who believe these words literally and act upon them—these are important instructions to the devout. In my observations, producers, whether devout Christians or not, believe these ideas. Gay or straight; Buddhist or Jew; mystical or rational: if people work with their hands, this is what they believe. A typical reaction was "Oh! *that* idea is from the Bible. I didn't know that. Oh well—who cares where it comes from—it's a good idea." These are definitive core values. Producers might not believe in Virgin births or redemption through capital punishment—the *taught* core values of Christianity, but they *do* believe in family owned and operated farms, among other passionately held ideas.

2 This quote, attributed to the radical Anglican clergyman Alan Watts, applies not only to Christianity, but to almost any religion. The irony of religion is that virtually all of the major religions were begun by antireligious types. It took Christianity three centuries to go from Jesus as Jesus to Jesus as God. Confucius, another antireligious type, was a deity in five centuries.

3 Roland H. Bainton, *Here I Stand*, Abigdon Press (1955).

There are hundreds of books about the impact of Martin Luther and the Protestant Reformation on history. Many are very good. What makes *Here I Stand* more interesting than most is that it is written by a true believer—it is the official version of the history of Luther for Lutherans.

The link between Lutheran cultures and advanced industrialization is a recent, mostly twentieth century phenomena and is not as important as the link between the Quakers and the industrial revolution in England. But it is very interesting because the technological achievements of nominally Lutheran countries are outstanding and all out of proportion to size of their populations. Sweden, with 8 million people, has roughly the same qualitative industrial capabilities (in some cases greater) as the United States with 250 million and could simply bury England in any meaningful industrial comparison.

It should be remembered that Christianity has not been such a force for progress because it was so strong but because it was so *weak*. One of the criti-

NOTES ON CHAPTER ONE

cisms directed at the Protestant Reformation was that it was going to bring on religious (and other forms of) anarchy. This is clearly did. There has never been such a powerful argument for at least a little bit of anarchy.

4 Alexander Hamilton, et al. *Federalist Papers.*

In many ways, these papers are more important than the Constitution of the United States. They clearly outline the class lines and power arrangements that would operate in the United States social system. Strangely, the *Federalist Papers* are almost never taught. When the operating blueprint for a society is not taught and the Constitution is taught but only honored occasionally, it is not surprising that citizens are confused.

5 James Burke, *Connections*, Little, Brown, and Co., Boston (1978).

Burke is British Broadcasting's (BBC-TV) science correspondent, who covered the American moon shots—among other significant efforts. His series about the cultural links between technological development and religion, geography, weather, economics, etc. are the finest I have found. It is *his* assertion that the Quakers are responsible for the industrial revolution and he provides ample evidence.

6 Jacob Bronowski, *The Ascent of Man*, Little, Brown, and Co., Boston (1975), pg. 274.

7 Thorstein Veblen,
The Theory of the Leisure Class (1899),
The Instinct of Workmanship and the State of the Industrial Arts (1914),
Imperial Germany and the Industrial Revolution (1915),
The Vested Interests and the Common Man (1920),
The Engineers and the Price System (1921).

It is impossible to summarize the great writings of Veblen in a few words. I chose to use the ideas from *Imperial Germany* only because it is most relevant to the times—modern reunited Germany is close to what Veblen hoped for in 1915.

Veblen is really America's only economic philosopher and is widely misunderstood. *The Theory of the Leisure Class* is a book that should be read first—it is the easiest and most accessible. It is more accurate now than when it was written. It took Veblen seven years of study and thought to come up with his masterpiece. It is not only stunningly perceptive, it is distinctively American. It may be Veblen's easiest work but is not easy to read. It was written in an arcane style even for his day. In 1899, people read and attended lectures instead of watching television. The real industrial expansion in the United States was only 30 years old and people had seen it happen. Veblen assumes all

NOTES ON CHAPTER ONE

this. A reader who thinks that milk comes from a store may find his writings confusing. To make life even more difficult, Veblen employs a specialized vocabulary.

With a dictionary, some third party commentary— especially Robert Heilbroner's *The Worldly Philosophers*—and persistence, *The Theory of the Leisure Class* can be read and enjoyed. It is worth every second of effort. John Kenneth Galbraith's explanation of Veblen in *The Age of Uncertainty* also provides an excellent introduction.

The Theory of the Leisure Class is probably not Veblen's most important book, that title belongs to *The Engineers and the Price System. The Instinct of Workmanship and the State of the Industrial Arts* influenced chapter four of this book and is my personal favorite. *The Vested Interests and the Common Man* is important for those who would understand the political passions of the 1920s—an era not terribly dissimilar from the 1990s.

8 Jean Ziegler, *Switzerland Exposed*, Allison and Bushby, New York (1981), pg. 46.

Anyone interested in understanding the reality of commercial Calvinism must read this book. John Calvin was the great Swiss reformer and though today, non-Calvinists are tolerated in Switzerland rather than burned, as was true when Calvin was alive, secular Calvinism is alive and well and dominates the Swiss cultural landscape. Accordingly, pure Swiss Calvinism is manifest in its most famous institution—its banks.

When all accounts are balanced, the Swiss economy is dominated by the banks. All other Swiss commercial ventures from cuckoo clocks to pharmaceuticals are bit players by comparison. Defining Swiss banking are the famous bank secrecy laws. The main reason for secrecy is criminal intent. From corrupt third-world dictators to arms dealers, Swiss banks enable most of the world's truly gruesome behavior. In the chapter "A Nation of Fences" (pp.39-66) it is argued that Switzerland's defining commercial activity is trafficking in stolen merchandise. Forget chocolates.

In one very funny story, a Swiss picture is drawn of the Fall of Saigon. Americans remember the pictures of frantic people fighting to get aboard the last helicopters leaving the American embassy. For the Swiss, it is the story of the controversy surrounding a Swissair jetliner which was asked to transport sixteen tons of gold looted by that great "democrat" Theiu—the same charming fellow for whose rule thousands of Americans died. How much gold made it to those famous Swiss vaults remains a question—that is a secret of course, but the answer is—it was as much as a DC-8 could carry.

For those who do not understand the Swiss economy, much of the reality of her banking is shocking. Most people would rather know the Swiss as those charming people who work for international cooperation—I know I would! Yet there is no denying the contention of his book that secular Swiss Calvinism has an ugly and very dark side.

NOTES ON CHAPTER ONE

9 It is fashionable in realpolitik-speak these days to discuss the subtle differences between the Sunni and Shiite Moslems. There is a valid reason for this kind of inquiry. Of the available ways to look at the world, the ones people hold most dear-beliefs held most strongly-can be called their religious beliefs—even if those beliefs are rational or atheistic. It is the set of beliefs that people will live and die for. Folks inclined to mysticism tend to hold to the religions with names—Catholic, Methodist, Lutheran, Shinto, etc.

Folks who insist on rational explanations for phenomena invent secular forms of their old beliefs. Psychology has been kiddingly called "secular Judaism,"—Sigmund Freud's attempt to cope with Catholic Vienna. Like the Catholics, Freud believed that dreams have meaning, sex is dangerous and leads to failed behavior the Catholics call sin, and that it is healthy to seek advice on absolution and restitution from someone trained in human behavior. For Catholics, the confessor studies in a seminary—for shrinks, it is science and the university. Not surprisingly, Freudian Jews and post-Vatican II Catholics coexist about as gracefully as is possible to the human condition.

The colossal outbreaks of twentieth century barbarism can be called the Great Protestant Civil War—a battle between the *secular* industrial manifestations of Lutheran Germany and Calvinist England. The protestant reformation in England was not especially religious—unless Henry VIII's desire for a new sex partner qualifies as a religion. Today, the state church of England is little more than an official art critic—the definition of *noblesse oblige*. England's reformation did not become *religiously* defined until the Puritans and the rule of Oliver Cromwell. Calvinism flowed into the religious vacuum created by Henry VIII's nonreligious "reformation."

The cultural tension created by a state church that is still not very religious and the more passionate manifestations of Calvinism, limits the possibilities of religious extremism in England. Though it can be argued that Calvinism defines English cultural norms, it is nowhere nearly as purely expressed as it is in Switzerland—or for that matter, in the United States.

The Anglophiles of the United State are from Puritan stock and so tend to exaggerate the influences of English Calvinism. England may only be 51 percent culturally Calvinist in reality, but from the American Puritan perspective—the dominant one on the Atlantic Coast—England appears to be about 95 percent Calvinist. To Veblen, who grew up in a society where farmers and tradesmen were Scandinavian and German immigrants while the towns and banks were owned by New England Puritans, the figure must have appeared to be about 99-100 percent.

The Protestant Reformation in Germany happened in reverse order. It was a purely religious and cultural movement co-opted by commercial and economic interests. Johan Gutenburg was the very embodiment of the secular values that defined Luther's cultural values. In modern Germany, Luther is not much remembered as the founder of a religious movement but as the person who de-

NOTES ON CHAPTER ONE

fined the German language—his ultimate contribution to the business of printing. German industrialists can trace their core values to Gutenburg.

The Protestant Reformation was not pretty. Lutheran movements became militarized. Sweden, full of Viking tendencies for a good fight, saved Luther's religion in the Thirty Year's War, but Germany was destroyed for 150 years. Even so, the cultural values of commercial Lutheranism survived—there was a version of Lutheran thought that spoke to the people who worked. Work is important.

The teaching is "Everyone must do his best." The agreement was that *"If I the worker give you the factory owner an honest day's work, then it is your responsibility as an owner to protect my job, see to it that I stay healthy and provide for me when I am sick and old, and pay me well enough so that I may marry and raise healthy children."*

This is the grand bargain of Krupp's *Generalregulativ.* (See also note from chapter four on Krupp's General Regulations) It became the heart of Bismarck's social legislation. It is the industrial labor bargain struck to prevent the repeat of the Revolutions of 1848 or 1870. The bargain has held up remarkably well in large measure because it is the Lutheran definition of social order. None of this would mean anything unless the bargain was fair and led to general prosperity. For producing a successful industrial order, it would be difficult to top the German technological prowess. German products are lusted after the world over.

The bargain defused the Socialist challenge which reawakened in 1870. Bismarck's social legislation of 1889 absorbed most of the Socialism's best issues. Medical care became the right of a citizen, old age was protected, children could not work in factories, and the owners would see to it that labor had the best working conditions affordable while still investing in the company. The political mutation of socialism to Social Democrats transformed them into secular Lutherans. In Lutheran Sweden, Lutheran clergymen are part of the government. The line between Social Democrats and Lutherans is visible only to an insider. Social Democracy is the Lutheran-Bismarckian compromise for the worker.

England lacked a religious buffer for worker's rights. Calvinist doctrines of predestination have the effect of telling a person that God shows who he loves by showering him with the blessing of riches. If you are poor, God must be preparing you for the really bad news. Perverting the doctrine further, secular Calvinists believe that being rich and owning land entitles you to make all decisions. Being rich made a person wise as God's agent on earth—other opinions were not required.

English factories were hell-holes. Towns were filthy. Disease ran rampant. Children worked 16-hour days in the mills and ate so badly that growth was severely stunted. Machines were dangerous and if a worker was hurt, he could starve. Calvinism's response to worker unrest was to blame demon rum. As a result, Methodism was little more than a temperance movement.

NOTES ON CHAPTER ONE

In to this mess strode Marx, an escapee of German repression. The Germans threw him out for arguing too vociferously for the worker portion of the grand bargain. He looked at English working conditions and came to the conclusion that the only solution was to seize the ownership of factories from the barbarians who created them—a prescription that had gotten him booted from Germany and now frightened the rentier class. To their everlasting credit, the English tolerated Marx, but the Marxist influence made labor relations in England much more militant.

Eventually English Marxism would mutate into the Labour Party—a party whose program became similar to that of the German and Scandinavian Social Democrats. Revolution became a figure of speech as working conditions improved following World War I. Yet, the Labour Party's prescriptions for social justice have always been more Marxist than the German grand bargain—which English Marxists have always viewed as a sell-out of principle.

Working-class Lutheranism had adjusted its expectations to the new reality of industrialism far better than their Calvinist comrades in England. Even so, they could not control the impulses of the upper classes.

The insanity of World War I has no higher justification than a juvenile pushing match. Germany was the new kid on the industrial block. She reasoned that her industry had become better than England's. Krupp makes better guns than Vickers, Nah! Nah! Na Na-Nah!

The German and English royal families were related and interbred down to the level where the definition of an insult was an important issue in life. Royalty's toys were their armies. "If I say my chaps can lick your chaps and you say no, the only way to find out is a sporting proposition—so let us have at it. Oh and, bye the bye, let us have a wager of, let us say, my empire against yours?" And so a generation of men were slaughtered over a wager. Fear, anger, and hatred would create an encore as soon as a new generation of boys could become soldiers.

Militaristic, upper-class Calvinism would define imperialism. Colonialism's perfect logic would dictate that if a person stole a whole country he is rich, ergo God must really love him and we should give him a title and bow to him. There were no checks on colonial mayhem, anyone thwarting God's plans should be exterminated. Cecil Rhodes, an upright preacher's son, would kill thousands of innocent people in the process of stealing Rhodesia. His plundered wealth would endow scholarships for the advancement of the ideals of English civilization.

In part because of secular Lutheranism, Germany never mastered the art of grand imperial theft—not that she did not try. Her failure is best attributed to the fact that England had all the imperial prizes before Germany could learn the imperial pretensions.

English secular Calvinism is individualistic: the high priests are the economists and bankers. German secular Lutheranism is directed to the possibilities of the group: their high priests are the sociologists and industrial-

NOTES ON CHAPTER ONE

ists. In either case, the high priests determine the social direction by defining the operating assumptions.

Until the onset of World War I, Germany appeared to offer Americans the superior model for industrialization. When North Dakotans named their capital in 1889, they chose Bismarck so as to attract investment and industrial expertise from Germany. The middle of the United States was settled by Germans who escaped the repression following the Revolution of 1848. Wisconsin became a state in 1848 and the political landscape has been dominated by the descendants of some of the best thinkers of German socialism—it is no accident that C. Wright Mills, America's best known sociologist, is part of the history of the University of Wisconsin. They were joined by Scandinavians who shared the German belief in the grand social bargain. The folks who owned the banks and railroads did not. The battle between the cultures was bitter.

With the coming of World War I, the progressive outbreak was crushed in the Midwest. The teaching of German was outlawed in schools and the members of the political organizations agitating for an American grand bargain were jailed for sedition. The horror of World War II would do more than end the practice of naming American cities after German ones.

German social ideals became badly stained. Fascism tore a gaping hole in the hull of German idealism—exposing a need for careful limits. Even the high-minded notions of shared provision, the strength of the group, and virtue of doing one's best, could turn mutant. The grand bargain assumes that everyone will do his or her best work, therefore questioning another's activities is a disrupting insult. I don't tell you how to do your job and you don't tell me how to do mine—is an important sentiment of the grand bargain. No one questions it because it usually produces a tranquil social order.

Hitler, who had lived with the working class and understood their commitment to the grand bargain, played this sentiment like a drum. Care for the group was perverted into militarism: shared provision—heroic sacrifice: quality work—superior weapons: intolerance to disruptions in the social order—efficient sanitation. Death camps were sold as a public health benefit catering to the German need for cleanliness. No one was to question decisions—no one was supposed to ask the questions. Those few Lutheran clergymen who dared to question the morality of mass extermination as a version of the grand bargain, were themselves killed. If German industry needed the resources of Russia and the Ukraine, then the most efficient methods of seizure should be used. If death camps needed an excellent way to kill people, the chemical industry would be happy to provide the necessary material that would meet all design specifications.

German-Americans, like all other civilized creatures, reacted with outrage at the value perversion. These people came from radical German utopian socialists. Many were pacifists. To this day, progressive American Germans react to Hitler's monstrous mutation of the grand bargain with stunned silence. Their silent cultural vote for the grand bargain is buying a German car. Otherwise, no one wants to talk about it.

NOTES ON CHAPTER ONE

Though invisible in modern American culture, the grand bargain lives on under other names. The Germans have renamed it as well. Today, it is most often described as German corporate culture. A German company like Mercedes-Benz will brag that it survived two wars, bombings, military occupation, marketing to an angry and traumatized world, runaway inflation, and if the speaker is young, Hitler. Mercedes is making more people happy with their production in 1992 than in 1942. The moral cover is, "Foreign policy is not my department, nor is history. My job has always been the same—make the best possible product."

In over forty-five years, secular German Lutherans have proved quite capable of preventing another collective insanity. In the process, they have confronted major moral dilemmas. Occasionally, this still turns violent. The German Red Army Faction, composed of otherwise serious young scholars—many Lutheran preachers' children, have been labeled terrorists for assassinating the barely-changed Nazis who retook control of some industries after World War II.

For whatever the cultural problems of Germany's past, recent history has shown that the grand bargain is still a powerful and useful industrial strategy. Germany's industrial equal, Japan, uses a modified version of the grand bargain with great success. In Scandinavia, where militarism died long ago and Lutheranism is the state religion, it exists in its purest form. If secular Lutheranism could be called "Gutenburgism," the Scandinavian version proves that their religious manifestation of a social bargain has historically very little for which to apologize. Few societies have less innocent blood on their hands.

10 Connie Bruck, *The Predators Ball,* Simon and Schuster (1988).
James B. Stewart, *Den of Thieves,* Simon and Schuster (1991).

I really *had* written about the producers and *predators* before *Predators Ball* was published. I liked the alliteration. Never has an idea been so thoroughly confirmed, however. There will be scores and probably hundreds of books written about 1980s Wall Street corruption.

11 Pehr Gyllennämmar, *People at Work,* Addison Wesley Publishing Co. (1977).

There is no better description of Swedish commercial values than those shown by the man who changed the nature of work at Volvo. There is a reason why Volvo workers are treated as well as they are, and why anyone who worries about worker exploitation should buy a Volvo to avoid guilt. Gyllenhammar caused the assembly line to be reinvented for health and safety reasons—an outstanding example of enlightened management.

12 Edwards Deming books are about the arcane world of quality control using statistics. The public Deming is much more interesting. His most memorable

NOTES ON CHAPTER ONE

social remark may be "The *only* way we will ever compete with the Japanese is to send them all the graduates of the Harvard and Stanford business schools."

13 William Manchester, *America's Caesar,* Little, Brown, and Co. (1978).

Lovers of history love the writing of Manchester. He is a very hard working writer who researches his topic thoroughly, analyzes carefully, and writes beautifully. *America's Caesar* could easily be his best work.

Manchester fought in a war run by MacArthur--reason enough to take the man very seriously. For the purposes of this book, chapter eight, *The Last Post,* devoted to MacArthur's duties as Military Governor of Japan from 1945-50 is most interesting. If MacArthur could only see the industrial giant he created when he disarmed Japan, he would be astonished and/or proud. Most likely he would be frightened!

14 The Japanese are beginning to understand their cultural power and are proclaiming it to the rest of the world. Shintaro Ishihara, a man characterized by the American press as "right-wing" in his now famous *The Japan That Can Say No!* (Simon and Schuster, New York 1991, pp. 80-81) is quite certain that Japanese culture has things to teach others.

> Underlying Caucasian racial prejudice is their intense class consciousness, a bias against people of the same race or ethnic group but of different social strata. The European nobility despised commoners and the lower social orders just because they were not of their privileged level, while the hoi polloi both hated the nobility and aspired to their prestige and social standing. Eventually, a democratic fiction that everyone is created equal obscured the obvious hostility between the upper, middle, and lower classes. The nobility prided themselves on a life of ease. Gentlemen did not go into commerce, much less work with their hands. Disregard the fact that they also benefited from the toil of the masses, the aristocracy viewed the other classes with contempt simply because they worked.
>
> This class consciousness has persisted into the modern era. Western societies have extraordinary disparities between strata and there is ubiquitous discrimination against the working class. In the United States, for example, fast-track members of the corporate elite will not even type a letter or do secretarial tasks for themselves. To go into the factory and get dirty and sweaty learning how products are made is beneath them.
>
> Class background largely determines the quality of education an American receives. Highly trained U.S. top management do not ask blue-collar personnel for suggestions about how to improve factory operations. Even of they did, the worker would probably have little to say.

CHAPTER TWO
MARXISM: THE FAILED PROMISE

PERHAPS, WITH THE COLLAPSE OF MARXISM in the late twentieth century, it would be wise to let the dead bury their dead; but Marxist thought cannot be dismissed so easily—even in defeat. Too many bloody battles were fought in the name of the working classes by Marxists. Too much effort was spent trying to defeat these movements. In the end, none of it mattered.

For many, Marxism was the only alternative to capitalism. For those concerned with the plight of the working classes, there is a feeling of hopelessness that the problems of capitalist excess will now go unchecked. Fortunately, capitalism has as many variants as countries which claim to practice it—many quite enlightened. As Hazel Henderson points out in her *Politics of the Solar Age,* capitalism and Marxism were merely peripheral arguments over how industrialism should be organized. [1] Marxism may have lost the arguments, but the contradictions of industrialism remain. Examining why Marxist ideas ultimately lost sheds considerable light on which of the remaining industrial forms should ultimately prevail.

Marx identified the predatory nature of the capitalism of his day and the resulting plight of the producers. His biggest analytical error was his failure to account for stratification within producer ranks. His descriptions of factory life are like the description of warfare from the point of view of a foot soldier. Life at the bottom has always been miserable no matter what the job at hand. He never describes the working characteristics of the people who *designed* the factories. In his defense, it would have been nearly impossible for him to do so because when he was writing, those persons formed a group so small as to be nearly invisible. It never occurred to him that the artisans thrown out of work in agriculture and crafts would respond by creating an industrial explosion.

When he wrote his *Communist Manifesto* in 1848, the experiments that would lead to the chemical industry were almost a decade off. There were no aluminum, petroleum, automobile, railroad, self-propelled farm machinery, electronics, or pharmaceutical industries either. Each of these industries has developed different characteristics based largely on task. Marx's description of capitalism is a description of mining, the textile industries, and other primitive industry. His descriptions of these industries barely need updating—even today, but it is a serious mistake to think that these industries are the models for all industry.

Marx himself was not a producer in the traditional sense of the word. Certainly he produced writings, but he built nothing, grew no food, nor wove any cloth. This distinction may have led to the fatal flaw in his reasoning: he recommended violent revolutions. Producer "revolution" may be an oxymoron. To be a successful revolutionary, one must first be a successful militarist. To be a militarist, one must abandon—a least temporarily—the life of a producer and become a destroyer. Once the revolutionary decision that killing another human for political reasons has been justified, the struggle enters a phase where producers have little to gain and much to lose.

ONE. Declaring war on the predators plays right into their hands. The predators are professionals at combat—it is what they do best. Moreover, whatever restraint a predator may show in times of peace is eliminated in time of war.

TWO. The persons most likely to head a revolution are the same persons who have the strongest military mindset. Although the foot-soldiers of a revolution can go home to their work when it is over, those who stay behind to govern believe in militarism as thoroughly as those they overthrew. They have become predators—even if they were not to begin with.

THREE. Winning revolutionaries kill the wrong people. Because excellence is a form of deviance, the producers most needed to build a successful society stand out, draw attention to themselves, and become targets.

FOUR. Revolutions breed counterrevolutions. This can keep predator-militarists in power for decades.

This pretty well sums up the problems of Leninist-Stalinist Soviet Union. The revolutionaries and those who repulsed the counterrevolutions (including World War II) were military men who confused building a country with running an army.

This does not mean all Marx's thinking was flawed. Marx believed that societies advanced from feudalism to capitalism to socialism to communism. A century later we can see that Marx's predictions of the nature of this progression may have been correct. What Marx failed to realize was that a *building* industrial society would probably contain variations of all four social organizations *simultaneously*. Moreover, some enterprises would stay at one level, perhaps permanently, while others would advance to the communist stage very rapidly.

Marx confused the growth of industrialization with historical determinism. He correctly noted that as enterprise grew in size, the social organizations necessary to operate enterprise became more collective.

One hundred fifty years of industrialization have taught that not all enterprise can or should become large. Marxism failed over this misunderstanding. In the old Soviet Union, all problem-solving, small or large, would follow a strict format that assumed collective decision-making was always superior. In fact, it was not.

Though often not pretty, the Soviet Union successfully organized and executed huge, mind-numbing mass movements of people, and forced large-scale industrialization, the defeat of Germany's best-equipped armies, and the building of transportation links covering eleven time zones. Large projects did not doom Marxism—small enterprise was the problem.

Eliminate the excesses of feudalism and it serves the organizational needs of agriculture well. Organized as it has been throughout most of the twentieth century, regulated capitalism serves the organizational needs of small and medium-sized producers. As projects grow larger they become more socialist in organization. Publicly held corporations coexist quite gracefully with the Socialist governments of Western Europe.

Examples of all four social organizations can be found in some mutant form in the United States. Owner-operated family farms combined with cooperatives form an interesting enlightened feudalism-socialism combination. Capitalism is still widely practiced and is particularly effective at bringing new products to market.

Larger social projects that require greater collective action pose difficult questions for a society dedicated to individualism as in the United States. The lack of a socialized perspective means delivery of medical care is absurd. Education has become a disaster area.

Large projects that form natural monopolies must be regulated. From the standpoint of social organization, publicly owned industries and regulated industrial monopolies are similar. Though there are many industries that would benefit from a more collective form of ownership, ideology prevents Americans from putting the necessary energy and funds into effective collective action. When it comes industrial cooperation, the keystone of industrial dominance, Americans chanted "central planning is impossible" like a mantra while the Japanese embraced the needs of central planning and made that work as well.

The ideological battles that accompanied the Cold War distorted Marx's ideas on both sides. The old Soviet Union, which believed that feudalism and capitalism could not be anything but evil, was systematically deprived of the advantages of family farm-cooperative agriculture and the innovativeness of industrial capitalism. The United States, believing that communism is evil, has been deprived of the possibilities for collective action while making competitive even that which is not—such as the national telephone system.

There even seems a sociology at work in what organization best suits each enterprise. People who grow things tend to be very traditional and family oriented. Producer capitalists tend to be dissatisfied innovators. Socialists tend to be caring humanists who make wonderful teachers and health-care workers. Communal projects such as power plants are best run by the most boring, bureaucratic, by-the-book person a society can produce. People may be predisposed by personality type toward each kind of organization. If this is true, the best-run society will have a slot for each type of person.

With all four social organizations in a given society, one merely inbeds an enterprise in the appropriate type of organization, and if things are not working well, moves it to one that works better. Let the farmers own the land they farm. Let the chefs run their cafes. Let curious experimenters and innovators guide the direction of industrial enterprise—only keep them out of places like Chernobyl which as the world found out, is not the sort of place for unauthorized experiments by dissatisfied industrial types.

If all four forms of social organization exist within a given country, then the debate shifts from which system provides *all* the answers to which system is the *best* answer for a certain application. If people are more or less free to choose which form of social organization best suits them, the abuses that have plagued each type might be eliminated. How abusive can capitalism be if there is a socialist alternative? How stifling can communism be if there is a capitalist alternative? The only possible losers in such a set-up are the ideologues.

Not all Marx's predictions were prescient. Marx did not see, nor could he have seen, how industry would shift from exploitation of humans to the exploitation of the biosphere. It is estimated that each person in North America uses the energy of 1,000 slaves in the form of fossil energy. These mechanical slaves make the lives of the modern working class better than the lives of the nobility 300 years ago.

Marx confused finance with industry. The role money plays in setting the industrial rules was understated. He concentrated his criticisms on the industrialists and ignored the financiers. Financiers may look and occasionally act like industrialists, but in truth the two are the soul of conflict. In large corporations today, the battle is waged between accounting and engineering. Those who win these battles determine the corporate face.

Marx called for ownership by the producing classes and, in many respects, companies owned or operated by engineers and scientists are the fulfillment of his prescription and are much better companies for which to work. In this respect, Marx was correct, though this was not what he had in mind.

Marx thought that widespread worker participation in management decisions would foster a better industrial state. Sometimes, as with the modern quality circles, this has clearly proven to be true. Because Marx failed to understand the importance of producer elites, he was unable to grasp the importance of industrial leadership and vision. "Designed by a committee" is a worthy form of slander because in some critical matters of production, individual genius counts for something and there may be no such thing as collective genius.

Marx postulated that labor is the basis of value but ignored the fact that all labor, even if equally strenuous, does not lead to the same end value. He taught that a capitalist is a person who owns the means of production and the workers should seize those means for their own good. As thousands of mine, factory, and farm owners throughout the industrial world can attest, merely owning the means of production means very, very little when the financiers decide that the industrial economy should come to a stop in an effort to defend the "value" of money.

Because of the profound changes in the nature of the industrial state since the time of Marx, his ideas from the modern perspective are not significantly different from those of any other member of the predator class. This is one reason that farmers and other members of the skilled occupations were seldom Marxists. An interesting comment to emerge from the Solidarity trade movement in Poland was a remark that "Capitalism is the exploitation of man by man—Marxism is simply the reverse."

Because Marx so significantly misunderstood the possibilities of the producing classes, his ideas have come to resemble religious teachings. Moreover, a significant number of those who called themselves Marxists treated his work religiously. Marxism failed because it could not organize work effectively; not for a lack of lofty goals. The collapse of Marxism may have marked the final irrelevance of neoreligious, environmentally insane, and technologically absurd social movements. Industrial problems are industrial problems and hewing to a Marxist ideology changes this little as the world found out at Chernobyl when a failed "communist" nuclear reactor did as much damage as any "capitalist" reactor would have done had it failed.

Although it is true that the quality of industrial output reflects the quality of the social arrangements, there is precious little evidence to show that Marxism in theory or in practice leads to a sophisticated form of industrial activity with less environmental damage. There is evidence that shows the reverse. When it comes to the most pressing problems of the twentieth century, Marx has little to offer because most are creations of the producer elites—people he did not even acknowledge. It is strange that there is much argument over minor conflicts but sadly most warfare in history has been between predator groups with trivial differences in ideology.

That, of course, is the legacy of Marxist revolutions. Because Marx taught that class warfare would occur between the powerful and the powerless, and failed to make clear distinctions between the different members of the producing classes, the Marxists revolutionaries were no more members of the producer classes than the rulers they replaced. As a result, a Marxist revolution meant only a rearrangement of the relationships in the hunter classes. From a producer point of view, there is no difference between a farm run by a landlord who lives in Paris and a planning committee in Moscow. In either case, the wrong people are making production decisions.

NOTE ON CHAPTER TWO

1 Hazel Henderson, *The Politics of the Solar Age*, Anchor Books (1981).

Anyone who suspects that economics is merely bad theology will love the work of Ms. Henderson. More than that, she exposes existing economic assumptions as environmentally preposterous. She is an environmental hero of the first order.

CHAPTER THREE
ARE PRODUCERS A CLASS?

WHAT IS THE PROPER WAY TO DEFINE A CLASS? First, prove the existence of a group with boundaries; and, two, explain what the members of the group have in common. The features of cultural commonalty will be explained in chapter four, but first there is this little problem of showing that producers constitute a class.

Class analysis is often based on income, but because a producer-predator analysis postulates that there are rich and poor members of producers *and* predators, the difference is fundamental yet esoteric.

Certain social scientists believe that membership in a class is determined by an individual's class awareness. This seems a reasonable requirement except the problems it creates in a country such as the United States where the notions of class and class interest have not been discussed in public for 40 years. Class notions are so unusual in American political discussions that the brief appearance of "class" politics in Jesse Jackson's 1988 campaign made even his supporters uncomfortable.

The producers' existence is validated by the common intuition. We know that producers exist because we can see what they have built. A building implies a builder. Similarly, we know predators exist because we can see people taking by force or fraud something that belongs to someone else. The main defining criteria is their differing strategies for survival. In a real sense, producers and predators are *occupational* rather than *monetary* definitions.

Establishing the existence of producers and predators is simple enough, but as we shall see, many people are not clear examples of either one. Modern social scientists classify most occupations of advanced industrialization as service occupations.

The existence of service occupations does not destroy the producer-predator duality. Service, after all, implies an allegiance to another person or agenda. For most of recorded history, "service" was merely another name for producers. Societies were pretty simple—there were those who ruled and those who served.

Because service often implies loyalty, it has become a tradition for those who would consider themselves a part of the service sector to identify with the ruling values of predators. Disruption to this social order did not occur until producing servants became economically important enough to forge a separate agenda and value set.

Figure : 3.1

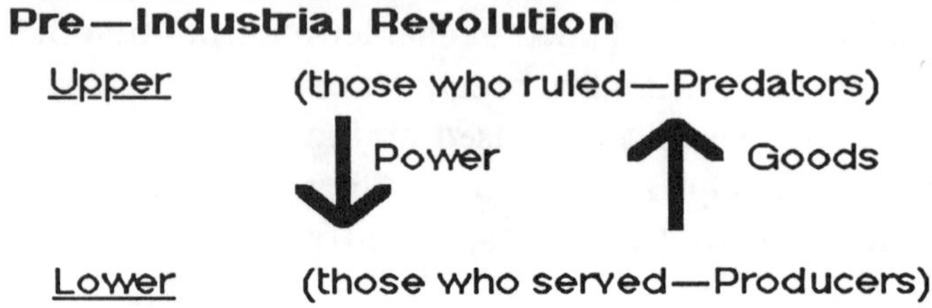

THERE WERE LAYERS of stratification within each group, but one thing was absolutely clear: the lowest member of the ruling classes was above the highest member of the serving classes.

The industrial revolution ended this neat arrangement. For the first time, producers achieved real power, but because industrialization, especially in England, was grafted onto feudal stock, the social arrangements were only slightly modified. Rich and powerful producers acted much like the worst of the old predators. This was the world as Marx described it.

Figure: 3.2

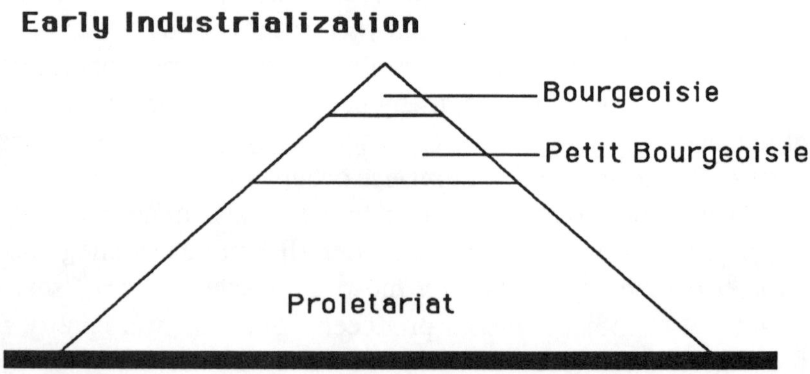

MARX'S PETIT BOURGEOISIE were the servants who did the dirty business of predation—not to be confused with the servants who cleaned the stalls. These latter were still lumped together with the producing peasants, builders, and mechanics to form the proletariat.

The late nineteenth-century American Populist writers were not as critical of business enterprise as Marx. Going into business, after all, was the main element of the American dream. Though there was not much evidence, the populists believed that producers could become successful and still maintain their producer attitudes. It was possible to reach the top without cheating anyone.

Figure: 3:3

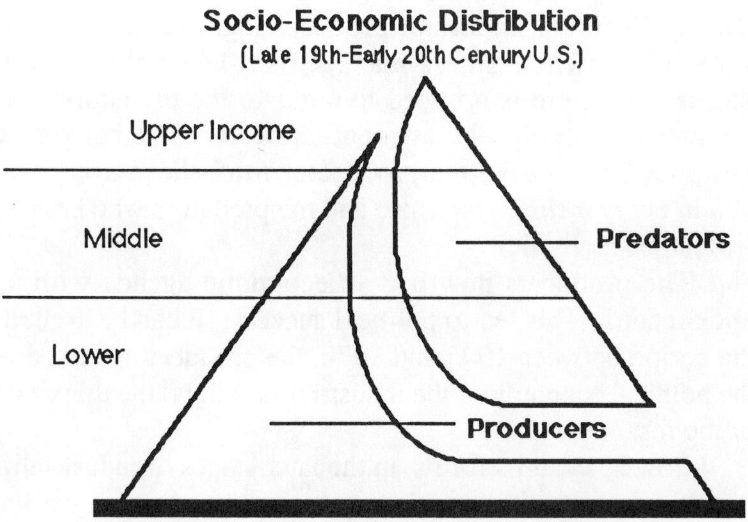

THIS GRAPH IS MEANT TO SHOW that, though a *few* producers had become rich, most were to be found at the bottom of the social order in any meaningful sense. The gap is meant to represent the emerging awareness of a separate agenda. Populists, Marxists, and progressives of all stripes agreed to and organized around the notions of difference.

The world got its first populist-producer billionaire in Henry Ford. Social progressives were enchanted. Ford seemed to have found the magic formula. He made his fortune *producing* something, paid his workers well, hired racial minorities, and embraced the 8-hour day. Better yet, when he made his fortune, he spent it on improving his product, opening a museum glorifying the history of the producing classes, and promoting causes like the end of World War I. John Reed, the American Marxist buried in the Kremlin wall, was convinced for a time that Ford's production theories and Marx's social theories would result in a Utopia.

By the same token, Ford's capitalist comrades were horrified. In spite of his incredible wealth and power, Ford was shunned by the wealthy and powerful for his ideas. As shown in the chart, the predators still held power and chose to make the 1920s miserable for both Ford and the class he championed. The idealism at Ford Motor lost its luster when the firm began to lose money.

In the end, Ford was to become a tightfisted, union-busting tyrant. Even so, important producer legacies remain: producers have their own business-management-leadership style that is successful—there is no need to emulate the predators; and, clean fortunes are possible—class conflict need not be between the rich and poor but between the producers who believe *everyone* can be rich in every *meaningful* sense and the predators who believe only a *few* can be really rich.

The producers now had an economic agenda with a proven track record. This led to political success. It can be argued that, in the period between 1945 and 1970, the producer agenda dominated the political economy of the industrial states and the thinking of most of the rest.

Power, in all its forms, in the later stages of industrialization is far too fragmented to enable one to state clearly that the producers are absolutely ascendant. (It seems as if most modern social science is dedicated to proving that no one has any power anymore. In a sense they are correct. No one seems to have real power because many persons and groups seem to have some. If a social scientist chooses to ignore the possibility of producer power, the fragmentation of power looks even more bewildering.)

In fact, a chart of power—both economic and political—might look something like this.

Figure 3.4

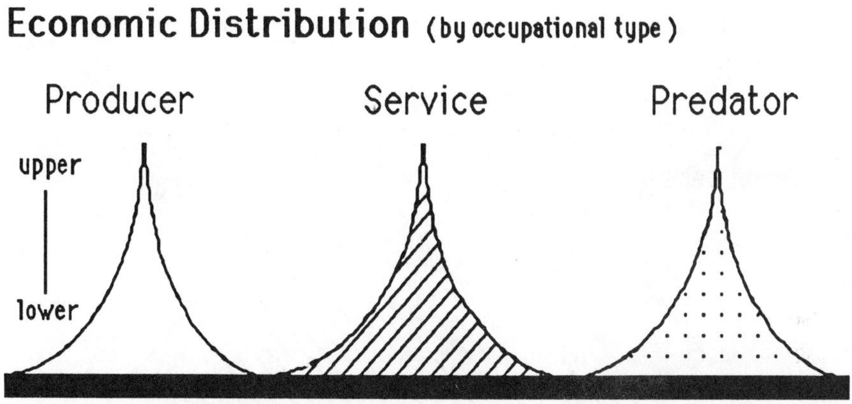

THIS GRAPH IS PROBABLY MISLEADING even if accurate. The newly significant service sector is, in reality, a convenient, but confusing, classification device because garbage collectors, bankers, and kings can be so classified.

With a huge service sector that can include bankers and royalty, there remain genuine predators. At the top are persons whose income is derived from ground rent or bonds, military rulers, and the like. At the bottom are the petty thieves.

There may be three basic occupational types but there remain only two agendas. Those who would serve are forced to choose between those who would produce and those who will not. It is possible for royalty and bankers, persons once considered the essence of predation, to serve the interests of producers. Kings and princes can be regularly seen pushing the products of their native countries worldwide. Bankers who live modest lives while promoting the economies of their communities are a regular fixture of the American midwest.

Figure: 3.5

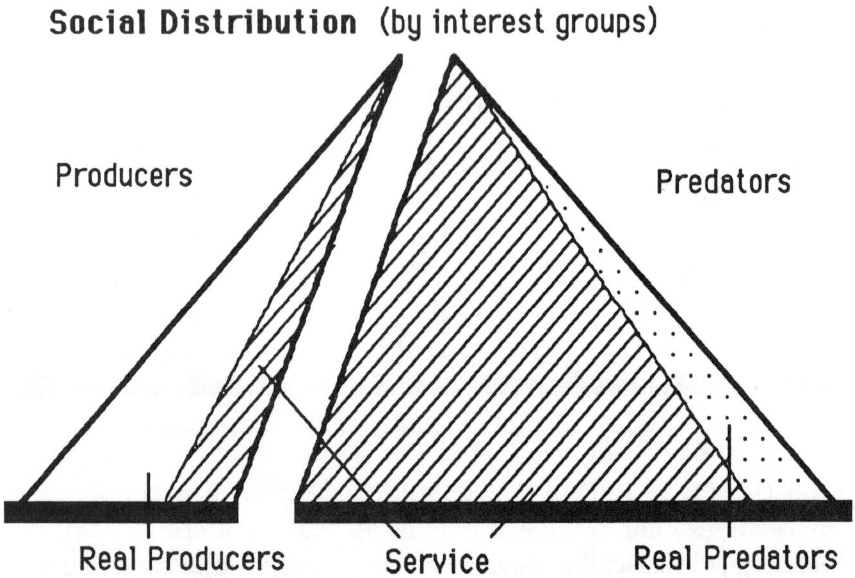

SUCH ARE EXCEPTIONS. Most bankers are predators, think as predators, and if not, serve the interests of predators. It is rare for any royalty (or anyone else living off inherited wealth for that matter) to justify income with real service.

An interesting picture emerges. It turns out that industrial societies do not need many real producers because they are so efficient. A single farmer can produce enough food to feed several hundred people. A punch press operator can make more parts in one year than he can consume in a century.

Producers find few *natural* allies in the service sector. The exception concerns those associated with industrial maintenance. The difference between building an automobile and repairing one is very small. If the goal is not merely an automobile but an automobile that runs, they are economically identical since an automobile that is inoperative has no (or negative) value.

Socially, maintenance and production people share an important similarity—both must understand and use tools. If the fundamental difference is between producers who use tools, and the predators who use weapons, then maintenance people are, in fact, Real Producers though they are usually classified as service workers.

If maintenance allies itself naturally with production, the rest of the service industries pose more problems for producer recruitment to their agenda. It seems the only certain way to create producers is to put tools in their hands and teach them to use them well.

Producers have appealed to the lower classes of the service sector through notions of class solidarity. This has not worked well. Trade unions look down on industrial unions and have joined forces only out of dire necessity. Producers can be awful snobs.

The upper classes of the service sector produce a whole different set of problems for producer recruitment. Because upper class members of the service sector are unlikely to use tools to produce anything, the predisposition of history is toward the predator agenda.

CLASS CONFLICT

THE RISE TO PROMINENCE AND POWER of the producing classes would tend to mitigate, one would suppose, the conflicts of class. In fact, something of the sort has happened. Serious scholars have portrayed the United States as a classless society. Ninety-five percent of the Japanese think of themselves as members of the middle class.

People do not talk of America as a classless society any longer, but class conflict has grown exceedingly complex since the issue was last raised. Any simplistic description of class conflict in terms of rich versus poor is probably doomed to fail because it is irrelevant.

The fact that both major interest groups contain upper, middle, and lower economic classes does not end battles between these groups but, in fact, provides a wider assortment of possibilities for conflict. The conflicts are of four major types: predator against producer, predator against predator, producer against producer, and producer against predator.

One: Predator ➔ Producer

THIS IS THE OLDEST CONFLICT. One who does not produce food and shelter for oneself must get someone else to do it. The predators have been extremely inventive over the years. Their methods have included slavery, imperialism, usury, ground rents, tithes, and taxation.

Of course, since the very upper predator crust does not do anything productive at all, servants who share the predator mentality have always surrounded them to do the actual work of profit taking, tax gathering, and rent collection. The real work of predation has been done by sheriffs, IRS agents, lawyers, judges, and an army of bureaucrats. Overseeing all this activity is the clergy (or other moral leaders) whose job it is to see that everyone agrees this is the best possible arrangement

Two: Predator ➔ Predator

THIS CONFLICT IS USUALLY CALLED WAR. History books are filled with the lurid accounts of these conflicts to which nothing can be added here. Within a given society, predator-predator conflicts are rare because loyalty is a big predator virtue while treason is a big predator sin. As a result, while tales of revolution and coups d'etat are common, historically they are quite rare. There are also recorded instances of bankers ruining kings, but these are even more rare.

Three: Producer ➔ Producer

THOUGH WIDELY MISUNDERSTOOD, producer-producer conflicts are common. They usually center around the issues of automation. The sophisticated tools associated with industrialization enable anyone with access to this tooling to copy exactly any product. The producer with the best original design and the best tooling will eliminate those producers with inferior products.

In the beginning of the industrial revolution, these producer-producer conflicts boiled over in social revolt as artisans were displaced by factories. The Luddite movement saw these displaced artisans smash sophisticated factory tools. The Luddite movement generated little sympathy. Few consumers were likely to complain about cheap factory-produced goods that were clearly superior to the more expensive goods produced by artisans. What really finished the Luddite impulse, however, was the realization that industrialization would also produce cheap, but sophisticated, tools. These tools would allow the small producer to fill the gaps in production left, deliberately or otherwise, by the large producer.

Small producers seek niches for their efforts for very good reason. Direct competition with a large, established producer is extremely difficult. A large producer has production experience, established ties to suppliers, known marketing outlets, and access to finance. Unless the technology of the small producer is far superior, there is no chance in a direct competition.

There are examples of new producers displacing old ones—such as when the $5 quartz crystal-microchip watch proved to be more accurate than the $5000 mechanical watch produced by the Swiss. Even today, the Swiss, with a four-hundred year head start in watchmaking, have not fully recovered from the competition of an upstart. Such examples are not rare.

Four: Producer ⟶ Predator

WHEN ONE THINKS OF ATTACKS OF PRODUCERS on predators, strikes, boycotts, and sabotage are what come to mind. And in fact, these are about the only options available to lower class producers.

Upper class producers have an option that they have frequently exercised: simply make things so very complicated that only those who made them know how they work. This has been the strategy of choice as producers have sought to increase their power. It has been highly effective. The world that producers have created by the end of the twentieth century is so complex that it is a rare predator who has even the vaguest notion how the world works.

Even the specialist servants of predation have a hard time understanding the smallest slice of the world they pretend to govern, regulate, or defraud. Producers go out of their way to make matters difficult (the most interesting little producer secret is that every extant process of production can be explained to any reasonably alert 9-year-old.) What makes a producer a genius is the ability to solve problems that have not been solved before. Once found, a good solution is "obvious." Even so, the predators and their servants exhibit an odd trait that makes this process of obfuscation easy.

Preservation of Archaic Traits

PREDATORS DO NOT KNOW much about producers (and their work) for an interesting reason that goes beyond the producer's tendency to make his work obscure. In many respects, predators do not know about the work of producers because they believe it to be beneath their dignity to know. They are *fashionably* ignorant.

The automobile provides a perfect example of a subject about which ignorance is quite fashionable indeed.

It is almost impossible to overstate the importance of the automobile on industrial society. It affects everything from city planning to sex. It is economically very important with millions of jobs at stake. Its impact on the environment in the form of resource depletion, air and water pollution, and the production of toxins is enormous.

Resource requirements affect international relationships. Rhodium, necessary for the production of catalytic converters used to fight air pollution, can only be found in commercial quantities in the Russia and South Africa. To fight air pollution, the United States has been forced to deal with one government or the other on a normal commercial basis. The choice of South Africa was for years the subject of a loud political debate and an object lesson in Cold War insanity.

One might assume that a subject of this import would demand widespread knowledge. In fact it has, but an odd phenomenon has occurred—those most likely to make major judgements on the future of the automobile are the ones with the least knowledge. Government transportation officials, environmentalists, car critics, and the like regularly make public pronouncements in which they mispronounce basic automotive terms, confuse facts, and generally give the impression they know absolutely *nothing* about the *real* automobiles that people drive.

More oddly, they seem genuinely pleased with their ignorance—treating it as if it were a badge of social stature. They even claim that their ignorance of the nuts and bolts of a subject allows them to arrive at more objective policy decisions. Now it is a fact that automotive policy decisions can be well formulated without knowing how a transmission operates. *It is also a fact that such people regularly make preposterous decisions because they do not understand the subject.*

Even if a person were to arrive at a policy-making group with genuine automotive knowledge, that person would be tempted to hide this knowledge for fear of being labeled a motorhead. In the august company where such policies are formed, the motorhead would be found guilty of a cultural crime—knowing what only the servants are supposed to know.

Producers know this social rule well. They have learned that scientific and technological issues are not to be discussed in polite company. As Veblen pointed out, the ability to do anything useful is suspect. Thorough knowledge of a subject demonstrates ability and experience, precisely the sorts of thing "important" members of society cannot have.

Ironically, people are not born fashionably ignorant—they pay good money to become that way. This cost is usually involved in getting a liberal education.

THE DANGER OF PRESERVED PREDATORY TRAITS

THE GREATEST PROBLEMS facing the industrial states do not stem from conflict, but rather from a rough sort of cooperation. Such cooperation is not voluntary but is rooted in the oldest of the power arrangements: predator power is cultural, economic, military, and political while producer power stems from a mastery over physical processes.

In spite of its seeming obsolescence, predator power is still very real. In the United States, predator values dominate the cultural forums. There are many reasons but one is significant. The National Security Act of 1947 put the United States on a perpetual wartime footing—a fact which dominates economic decisions, distorts political institutions, censors newspaper reporting, and muddles the educational processes. The Soviet Union, the putative object of this war, was forced by this action to choose a similar set of values which have caused similar problems. Because of their commitment to the predatory values of militarism, these two nations were often called superpowers. They have also been called the

"Klutzes of the North" by Gore Vidal for their persistent problems with production.

Unwilling to challenge predator power or its values directly, producers historically have found themselves in the unfortunate position of increasing the powers they seek to thwart. Producers have made war more deadly with their weapons, authoritarianism more pervasive with big-brother computers and surveillance technologies, and demagogues more influential with television. If this were not enough, the producer's failure to challenge the notions of wealth, monetary policy, and usury has multiplied the predatory aspects of their own industrial enterprise. To succeed under such assumptions, a producer is forced to violate nature even as he is exploited. As a result, the main crises of industrialization is the environmental crises.

The predators disavow any responsibility for their role in the industrial rape of the planet and their response is as old as history. Return, they say, to the Garden of Eden. Go backward to a time when the producers did not present such ghastly problems. Roll back the *social* gains of the producers and the problems they have caused will disappear.

There is no retreat from producer problems. Industrial-environmental problems have already been created and they would not vanish even if the producers and industrialization were to disappear overnight. There is no Garden of Eden solution. Producers are responsible for the magnitude of our environmental problems. Only they understand their scope. Only they can create a solution for problems that already exist.

The implications of this reality are enormous. It means that rather than a return to predator values that have so characterized the 1980s worldwide, producer influence must be extended further into the cultural, economic, and political arenas—arenas where producers traditionally have feared to tread.

CHAPTER FOUR

THE THEORY OF THE INDUSTRIAL CLASS, OR

WHEN CREATIVITY BECOMES IMPORTANT!

"The hand is the cutting edge of the mind." *Jacob Bronowski* [1]

WHEN VEBLEN wrote his book, *The Theory of the Leisure Class,* he assumed that his readers knew little about the values of the ruling class and used industrial values as a starting point. Today, because leisure class values dominate the mass media's features on politics, finance, militarism, travel, lifestyle, and manners, the problem is reversed. Because of this, any discussion of the producer class and its value systems seems almost foreign in a contemporary setting. Ignoring or misunderstanding the producer class, however, is a dangerous mistake.

Producers are producers because they create things. Any description of a good producer will eventually begin to sound like a description of an artist. If one subscribes to the notion that artists are simply creative people, then producers are artists even if they are called farmers, engineers, or factory workers.

THE PRODUCER AS AN ARTIST

FOR A GROUP that has been around for thousands of years and has grown to wield such power, the producer class is not well understood. The reasons for this lack of understanding are many but two stand out: People who make things seem not to have the resources, interest, or inclination to write about what they do; and, those who have developed specialized skills have a strong economic incentive to keep those skills a secret.

In many respects, it does not matter if the producers reveal their secrets. Valid judgements about a producer can be drawn from what is produced. Knitting a sweater or building a skyscraper may not seem to have much in common but necessary attitudes concerning planning, attention to detail, initiative, and perseverance are very much the same. The difference between small projects and large projects is one of scale, not of process. The problems of scale merely change organizational requirements.

In any field of human endeavor, some folks are naturally gifted. Natural gifts are quite visible in athletics and music. The same is true for builders. What absolute pitch or a great voice is to a musician, what quickness is to an athlete, the ability to conceptualize in three dimensions is to a builder. This is merely one of many necessary traits of the producers. There are others.

North America is a particularly good place to examine producers. The United States was founded by revolutionaries who subscribed to producing class values. Even more importantly, the United States was a giant construction project—there was need for many builders. They came from all over the world and left their legacies.

CREATIVITY AND PRODUCER CLASS HISTORY

OF ALL THE DISTINGUISHING CHARACTERISTICS of society's producers, the ability to create is the most important. Real creativity is comparatively rare and widely misunderstood. In the twentieth century industrial states, creativity is most commonly, and wrongly, associated with the fine arts. At one time, artists were certainly members of the producing classes with leading edge skills. What is common to the art of Bach, Michelangelo, and DaVinci is that they were first accomplished craftsmen. Their creativity grew from the fact that their skills were so phenomenal that they were required to invent projects to showcase their abilities. DaVinci's "Last Supper" was a demonstration of the newly emerging understanding of perspective which, in turn, was an offshoot of the latest theoretical understandings of the workings of light.

What passes for fine art in the twentieth century is not a showcase for scientific understanding or leading edge skills but a glorification of the primitive. Great casting techniques are employed by the builders of the B-1 bomber, not in the statuary in public places. Attention to detail is seen in the fitting of the heat tiles on the space shuttle, not in the splatterings of Jackson Pollack. In the very real sense that twentieth century artists would rather theorize about art than actually create, art has become a leisure class activity.

Producer class creativity is downgraded because it is functional. If a Boeing 747 could not fly and was more poorly constructed, it could easily pass for a work of art entitled "Aluminum Bird," and be given a suitable place in a museum of modern art. [2]

Producer creativity is different in another important respect; it is harder to ignore. An opinion of the music of Arnold Schoenburg is likely to be far less passionate than an opinion of the work of Robert Oppenheimer (the head of the Manhattan Project that produced the first atomic explosion.) When such a creative event takes place, the world is never quite the same again. All significant technological and social change is tied to the possibilities of producer creativity.

In this context, discoveries must be considered creative events. When Dmitri Ivanovich Mendeleev discovered the periodic table of elements—probably the single greatest accomplishment by one human in a lifetime—he was not creating something new. He was merely stating accurately what had always been true. It was up to his mind to formulate the right idea. This process of formulation is the creative act of discovery. This creative act of discovery is the essence of science.

Those who would learn science are taught the scientific method. These *rules of verification* are really the only part of science that can be *taught*. The scientific method is useful because it is important to have an agreed upon set of rules so that science can eventually discriminate between the truth-tellers and the charlatans, fakers, and those who are simply wrong.

The real science, the science that divides the great discoveries from the rest, is in the forming of the correct hypothesis. Until the correct hypothesis has been formed, all verification procedures and testing methods yield only the knowledge of what does not work. The ability to formulate the correct hypothesis—the essential, creative act of science—is not a subject that can be taught in school. It is not even an ability that can be described, even by those who possess it.

Ask a scientist how he or she came by an idea and you get a story. Albert Einstein claimed that he got his ideas about relativity while riding a streetcar. Another version has him dreaming of riding on a beam of light. Legend has it that Archimedes was taking a bath when the ideas about buoyancy came to him. The lightning strike of an idea that is immediately, apparently true, which appears in an instant and can stand the retesting of results for hundreds of years, is one of the unique happenings in the universe.

Scientific discovery, agro-industrial innovations, and inventiveness are the creative high water marks of the producers. These are the acts by which they make history. The invention of the printing press changed the world more than anything done by either Napoleon or Alexander the Great. Contained within the realms of producer creativity are most of the positive possibilities for the future. The predators can only promise one form of destruction or another. Such futures have always been available. Only the producers can *create* a future but such creativity has its dark side.

Producer creativity is haunted by its mortality. Creativity is a highly intoxicating addiction leading to hubris because creativity is considered Godlike in most cultures. Christians universally recite, "I believe in God the Father Almighty, *Creator* of heaven and earth."

In this respect, the idea of God is most telling in the origins of Freemasonry.[3] The Old Testament Jewish God was modeled after a male, warlike, tribal chieftain. This role was further emphasized by the Christian leaders of Europe's Dark Ages who found the model useful to the exercise of divine rule. The Freemasons, in contrast, found "God the Creator" more relevant to their lives than "God the Warrior Father."

Renaissance cathedral building was a tricky, dangerous, highly-skilled enterprise that required several generations to execute. The need to maintain the necessary body of construction knowledge over time fostered new methods of instruction. Training a new master builder took years. Because there was so much life-and-death information to memorize, the masons chose to use near-religious ritualised learning methods. The ritual taught that each newly acquired skill brought the apprentice closer to the Great Builder. Today, most Masons have little to do with building but their secret apprenticeship rituals remain and they still pray to "God the Creator" rather than "God the Father."

It is possible such thinking would better suit the modern human producer-creator. Scientific and technological creators must understand that their work is never properly measured against that of their colleagues, but against the forces that created the biosphere.

The people who gave the world nuclear power might better have asked themselves, "Do I really know enough about nature so that I should release the power of the sun on earth in order to boil water?"

Frankenstein's monsters come from a lack of humility in the creative process. Imitating "God the Creator" must be an exercise in restraint or the consequences are apocalyptic. If the producer-creators simply must play God, they need first understand that all creations decay and die. Mortality is the genius of the natural order. Everything in nature has an existing agent of decay.

Forgetting this, producers have made countless products with no natural form of decay. Human creativity is, by definition, a second-rate exercise until the apprentice creators understand that molecules are merely borrowed in nature, and that everything created outside the natural agents of decay must just as certainly be de-created some day.

Producer creativity is often very difficult to assess because producers have kept their creativity hidden behind veils of secrecy. Seemingly, producers are so enchanted with their creative powers that they are happy to let their work stand on its own merits. The need for fame, the driving force of predators, is strangely absent producers. The great cathedrals of Europe were built without anyone feeling the need to scratch one's initials somewhere. (The builders actually did leave their marks, but in subtle and unobvious ways.)

Needing fame themselves, predators, who must have other forms of attention because they cannot create, were more than willing to cooperate with the producers' unwillingness to take the limelight, and simply ignored them. The result is that when predators notice something new, the important creative work of the producer classes has been finished for a long time.

Producers sometimes find they cannot even communicate among themselves. When the lives of the producer giants of the past, such as Leonardo DaVinci, are examined, we see that their inner revelations of what was possible in the future became so real to them, that they found it impossible to understand their associates. DaVinci found that his mind was in a different century from the real world around him and withdrew into a life of almost total isolation.

Of course, there is a less esoteric reason for all this secrecy. The longer producers keep the predators out of the know, the longer they keep their wealth. The infatuation with high-tech enterprises stems from the fact that everyone knows, producers and predators alike, that all the big money is made when the enterprise is new. At one time, steelmaking, currently considered a declining industry, was *the* high-tech business. Since the predators can never build but only destroy, the game has always been to make the enterprise as successful as possible before the hunters show up and wreck things.

To get some idea of how far ahead the producers usually find themselves, a citation from Adam Smith should suffice. In his *Wealth of Nations,* Smith uses the example of the pin factory. He points out carefully how the work is organized into specialized tasks so that productivity goes up.

From the producer point of view, someone has already thought of this idea to break down the production of pins into individual tasks, he has convinced everyone from his backers to his workers that this is a workable scheme, he has reorganized the work-flow patterns according to his plans, and has made the idea work well enough so that it turned a profit.

After all the important decisions have been made and the pin factory is in operation, along comes Adam Smith, the prototype of the absent-minded professor. This representative of the ruling predator class drifts in out of his fog long enough to recognize a significant addition to the science of production and says, "now there's a good idea." By explaining this producer class invention to the ruling-predator class in terms they could understand, Adam Smith changed history.

By codifying one industrial paradigm, which while effective, is clearly limited in its application, Adam Smith gave the ruling predator class a good way of becoming very rich and staying very powerful. Unfortunately, this closed the door to alternate industrial schemes so that virtually all subsequent industrial development has been skewed by the set of principles that Adam Smith had to make simple enough for his employers to understand.

Unfortunately, this also means that industrial innovations, especially in organization, must face political and social as well as technological hurdles. Smith's pin factory is only a model for mass production—a fatal flaw in the twentieth century where very little of the necessary remaining production should be so organized.

SKILLS, PROCEDURES, AND VALUES—THE ART OF TECHNOLOGY

OF ALL THE PEOPLE who would study music, only a tiny handful ever become composers. Composition cannot be taught. Music theory can be taught. Most of the skills necessary to play a musical instrument can be taught, but to write music, a composer must hear a private sound in the ear of the mind. If the person who hears the sound in the mind can translate that sound into the agreed upon markings that constitute a musical score, then the rest of the world can hear the sound that once was only in the composer's mental ear. The ability to translate the sound in this ear into a musical score can be taught, but for there to be new music, there must be the sound. The sound is a gift.

The sound is not even dependant on actual hearing. Beethoven wrote his masterpiece Ninth Symphony while totally deaf. Since some slippage is inevitable between the sound of the mind and the sound of the orchestra and chorus, one can only imagine the incredible music that was in the head of the great Beethoven and take comfort from the fact that for him, the immortal Ninth was never tarnished by a missed downbeat or an E string gone flat.

Of course, it is not merely composers who must have gifts like the gift of the sound. An architect must have the gift of the eye. The architect who cannot look at a vacant lot and visualize a building, cannot design. The architect may draw beautifully, or make fine models, or take great photographs. Even with top grades in school, the architect who cannot look at a blank piece of paper and see the squiggles that will communicate the building in the mind's eye to those who must build it, will be, at best, only an illustrator.

Those who initiate the technology that becomes a part of the everyday lives of the members of the industrial states, must have the same sort of eye a creative architect has. The United States Patent Office recognizes the creative function of technology by referring to patents that have been registered in the past as "prior art." Before there was a drawing or a working model, there was a picture in the mind of the inventor.

The notion of the inventor as an artist is an extremely difficult one for the twentieth century mind to accept. Art and technology have become separate subjects. Of course, the distinction between art and technology is not only arbitrary and artificial, but a recent historic development. The same eye that makes a good artist can be used to make a good architect or inventor. Leonardo DaVinci was clearly all three. European industrial designers do a superb job of blurring the lines dividing artist, engineer, and inventor.

Art and technology share an important similarity. In order for either to be great, they must be seamlessly whole. One brush stroke could ruin the Mona Lisa. One wire astray can halt a $40,000 automobile. This similarity points out the greatest difference between the Mona Lisa and a Lexus. The perfection that is the Mona Lisa began and ended within the person of Leonardo DaVinci. The perfection that is a Lexus is not only not the product of one person. It is not even the product of the knowledge of one century. If technology is an art, it is a *cumulative* art in both a historical and a *cooperative* sense. While the art of the Mona Lisa could have happened in almost any historical period, the art of technology must have two important preconditions. To thrive, the art of technology must have an untrammeled flow of information and enlightened social conditions.

A vandal could destroy the Mona Lisa with a knife or a brush loaded with paint. A vandal could halt a Lexus with a wire cutter. Unfortunately, the art of technology that is embodied in the Lexus can be destroyed by a vandal wielding something more serious than a wire cutter. The art of technology can be driven from existence by censorship and political repression.

Censorship is a prime cause of technological backwardness. Necessary to inventiveness is knowledge of what else similar is happening in the rest of the world, and creative thoughts. Both feed on information. If an inventor does not know the state-of-the-art in a field, there is a very great risk of repeating someone else's work. This is reinventing the wheel—one of the great time-wasters.

Obtaining the information necessary for creative thoughts is a far more interesting and esoteric problem. Technological creativity is usually the result of examining a problem from a new perspective. Limiting the flow of information reduces the possible number of changes in perspective. As a result, any country that places any form of restriction on the free flow of information, of any kind, will be more technologically backward than those countries with less censorship.

GREAT ARTISTS MUST OFTEN WORK FOR PREDATORS

ONE AREA WHERE THE HUNTER AND PRODUCER CLASSES have cooperated throughout history is in the area of weapons making. Those who manufacture arms have generally been treated better than any other members of the producer class. Not only is this true today in the sense than those who work for General Dynamics, from the top to the bottom, live better than their counterparts working in a textile mill, but it probably has always been true. It may be safe to assume that the knight's armorer lived better than the field worker. Best of all, the outcome of a battle meant very little to the arms makers because a new boss was usually very much like the old boss.

The life of the arms maker combines the advantages of the hunter class lifestyle, the satisfyingly creative work of the producer class, and the unique benefit of never having to join the hunters in battle. Even when the battle turned against the Germans on the eastern front during World War II and manpower needs were serious, the rocketmakers at Peenemünde were exempt from service. While men their age were dying in some of the most brutal combat in history, the rocketeers would work in well equipped laboratories doing the kind of work they really wanted to do.

Since both the warrior and the arms makers are considered great patriots, there is a considerable advantage to being an arms maker, not the least of which is the pick of the widows when the battle is done. Keep this arrangement up for a couple of millenniums and we have the ultimate irony of the twentieth century: at least a score of people want to make and sell weapons for every person who actually wants to get into a fight. Nations that are technologically advanced enough to make weapons find it difficult to find real soldiers in their midst, and must often rely on technologically unsophisticated nations or mercenaries to use the weapons in combat.

Arms manufacture is not an ideal example of institutional influences on the direction of producer class creativity—but it is the best one around. It is very hard to tell who is leading whom around by the nose in the arms business. On one hand, without the predatory practices of the hunters, arms manufacture would lose its reason for existence. On the other hand, because the entrance requirements to the arms manufacturing business are so stiff, extremely bright people are involved who have it within their power to get pet projects approved.

Werner Von Braun understood the need to play to militarist superstitions perfectly. There exists little evidence to suggest that Von Braun wanted anything out of life save the resources to build rockets. The V-2 had no chance to change the course of World War II. Von Braun knew this better than anyone, but when there is a war on, it is best if work appears to be war-related.

As an American immigrant, Von Braun was able to continue his work. In one master stroke of fundraising, he introduced the concept to Washington of the "high ground" in space. Space—where the notions of up and down become meaningless—has no high ground. High ground was a concept the hunters gathered in Washington could understand. Not long after the high ground argument became official Washington dogma, Von Braun got a virtual blank check to build whatever rockets he and his colleagues could dream of.

In Von Braun we have a perfect example of producer power. He was able to get his agenda approved by appealing to hunters in their language. There are many would-be space explorers in America who want to build big, exciting rockets. Most occupy the high end of the intellectual spectrum. If they are not allowed to build civilian research rockets, they are forced to stampede the dimwits who believe there is a high ground in space into finding money for militarization. Von Braun may have employed producer power, but there was a hint of desperation about its use.

Even so, it is a fact that producers, through the power of the techniques used by Von Braun, may have become the ascendant class in the latter half of the twentieth century.

Is "Producer" or "Predator" a Personality Type?

"Football is to physical culture what bull-fighting is to agriculture"
—Thorstein Veblen

THE SPECULATION that people are born to be predators or producers is a trifle esoteric. Books have and will be written about the nature vs. nurture argument and it is not necessary to debate the validity of the evidence here. Even so, it is important to examine the implications of the outcome of this debate.

The nurture argument is simple. Producers and predators attend different institutions of higher learning that are not only located on separate campuses, but are often located in different cities. The separation of learning into institutes of technology and liberal arts colleges has over time created the distinction between the two types.

The nature argument is based on an article of faith that goes something like this. Deep inside every person is the real you. Nothing can change the real you; not privation, hardship, or catastrophe. In fact, stress is what brings out the real you. In the absence of stress, the real you will show itself by your avocations. What you do for a living may be a matter of expediency. What you do for enjoyment is the real you.

The implications of the nature-nurture debate as applied to producers and predators are interesting. It is possible for someone with a producer occupation to have a predator hobby—such as a carpenter who hunts ducks. A banker who builds model trains in his basement would be an example of a reverse situation.

Because the virtues and attributes of the great predators are well known, it is appropriate to examine the possibility that there is a real you of producers. The foremost thing to remember about producers is that they often fall in love with their work. This may be a fact born of evolutionary necessity. Since those who built and produced were often exploited, they compensated by learning to love the work itself rather than pursuing work as a means to some other reward.

Although predators talk about their careers, producers talk about their work. Work is so important to producers that they will often work for long periods of time without pay. Keeping the project funded is often the foremost consideration. From Mozart, who died of the complications of poverty, to the unknown modern inventor working for years without pay on a breakthrough he just knows will work, the pattern is the same. For the natural producer, everything can be sacrificed for the project. Leaving behind significant work is also a certain route to industrial immortality. These attitudes, of course, are the reason producers are so easily exploited.

They are also a certain refutation of the idea that coercion must be employed to get producers to work.

The Stonecutters, a recent Oscar-winning documentary, is a charming and accurate account of the traditional producer values of the cathedral builder. The stonecutters are very proud of their work and of their profession. "What we know about ancient history," one says, "we know because of carvers. It's the second oldest profession." After we are reminded that Michelangelo was a stone carver, the stonecutter adds that God, who carved the ten commandments on stone, could be considered a stonecutter. It is as if he is saying, "so you're a banker—name a banker with the stature of Michelangelo or God. Go ahead—I'm waiting."

In the final scene of the film, a retired stonecutter is standing in front of the Washington Cathedral. He says, "You're up on the scaffold; you swing your hammer; and a tiny chip flies out. You do that for forty years and you look up and discover that you have carved a lot of stone." The retired stonecutter turns and looks at the magnificence of the Cathedral and continues, "And when you look at what you've done, you realize that you haven't wasted your life." The intrinsic motivation of a builder has probably never been better said.

Stonecutters are not the only producers with a fascination for history, but producer fascination with history is different. The predator history of laws and kings and battles and conquests is somehow quite boring. Producer history is about discoveries, the development of tools, the rise of great industries, and the building of great transportation links.

Producers tend to be nonpolitical. To them, the notions of the nation-state have traditionally seemed faintly silly—to be a Freemason implied the freedom from the constraints of political borders. To the modern producer, for whom you work and how well your industry is doing is a greater determinant of how well you live than in which country you reside. It is more than a simple matter of loyalty to the firm eclipsing loyalty to a nation-state. Producers view the world in terms of suppliers and customers rather than allies and adversaries. Throughout the Cold War, American farmers complained loudly about the designation of the Soviet Union as an enemy. For the farmer, the so-called enemy was a prime customer.

American agriculture realized that the foreign policy establishment was far more harmful to their interests than the official enemy.

More interesting still, the industrial class has sought representation in all political parties. [4] The ability to create, manufacture, or otherwise produce something that was not there before, the real defining characteristic of producers, brings with it no definite operating political characteristics. [5] This fact most likely stems from the realization that politics, being the traditional province of the predators, has rarely involved itself with production problems.

Producers *hate* supervision, especially predator supervision. Inspection, the predator idea of quality control, is viewed by producers as outright harassment. Even the most basic industrial work, such a running a punch press, is usually a far more sophisticated job than the predators, who notice only the noise and the dirt, will allow. No one knows more about running a punch press than the person running the punch press. Recognition of this most obvious fact is at the heart of Japanese quality control. By teaching the industrial workers self-inspection, the Japanese have raised quality control to levels never before achieved in history, while abolishing inspectors and fix-it lines, the essential fixtures of predator quality control.

Producers believe most often in a strict meritocracy: persons should be judged by their abilities, efforts, and accomplishments. There is, in fact, a highly stratified producer pecking order. Placed along side of the more widely known leisure class pecking order, the industrial order would look something like this:

Industrial Class	Leisure Class
Nobel Prize Winners	Kings
	Presidents
	Heads of State
Theoretical Scientists	Political Advisors
Process Inventors	Military
High-tech Entrepreneurs	Financial Leaders
Process Engineers	Business Leaders
Industrial Designers	Elected Officials
Product Inventors	Lawyers
Development Specialists	Economists, Clergy

THE INDUSTRIAL CLASS even takes different parables from nature. A favorite children's story in the hunter-leisure class is the story of the grasshopper and the ant. The ant is a diligent *collector* who has provisions for the winter. The grasshopper consumes as it goes. Obviously, within the confines of the leisure class options, the ant is clearly the superior role model.

The industrial class is not so nearly infatuated with collecting. They would rather model their behavior after the beaver. Like them, the beavers work all the time: they are very industrious. The beavers are also more clever than the ants: they alter their environment so that there is a steady supply of food, obviating the need for collecting.

Collecting is a leisure-class diversion. Like most diversions, it fulfills a need to practice proficiencies, which may have been survival skills in the past, but are anachronisms in the twentieth century. The industrial class diversions, while fulfilling largely the same function, are clearly different. The interesting fact is that these diversions teach very different sets of values.

Industrial Class Diversions	Leisure Class Diversions
Team Sports	Team Sports
Yachting	Football, Soccer
Auto Racing	Basketball, Hockey
Individual Sports	Individual Sports
Flying	Jogging
Surfing	Swimming
Hobbies	Hobbies
Gardening	Hunting, Fishing
Model Building	Collecting

IN INDUSTRIAL STATES where leisure time is possible, such diversions have become major enterprises. Play is big business. Values expressed as diversions offer unique insights into the very real possibility that membership in the industrial or the leisure class is not a function of training or environment. Given more or less equal amounts of money for discretionary diversion, the person who joins a country club is clearly a different sort of person than one who builds airplanes in the basement.

When play becomes a spectator sport, there has been a tendency to make sport into something other than a diversion from work. Team sports are favored in the United States as much for their usefulness in transmitting predator values as for entertainment. Unfortunately, about the only skill that can be learned by playing football is how to refight World War I.

Once an examination of the personality differences between producers and predators is undertaken, many things besides occupations and diversions differentiate the two groups. Some were mentioned in chapter one such as giants, heroes and anthropological roots. Others will be explained in later chapters—especially those on economics; what follows is a chart of the most obvious differences.

HOMEOPATHIC DIFFERENCES BETWEEN PREDATORS AND PRODUCERS

PREDATORS	PRODUCERS
Anthropological Roots	
Hunting	Agriculture
Definition of Success	
Will have to do no work at all	Work will have its effects on millions
Means to Power	
Weapons	Tools
Violence	Mastery of physical processes
Religion	Increases in scientific knowledge
Manners	
Laws	
Favorite ways to Get Rich	
Slavery	Inventions
Ground rents	Mega-projects
Tithes	Military procurement fraud
Taxation	New businesses
Stock manipulation	Producer monopolies

PREDATORS	PRODUCERS
Means to Personal Success	
Who you know	What you know
Appearance	Ability
Heroes	
Alexander the Great	Thomas Jefferson
Erwin Rommel	Benjamin Franklin
Marshall Zhukov	Thomas Edison
J. P. Morgan	Henry Ford
Donald Trump	Alfred Nobel
Favorite Governments	
Royalty	Democracy
Dictatorships	Anarchy
Favorite Publications	
Forbes	Business Week [6]
Wall Street Journal	Inc. Magazine
Washington Post	Car and Driver
Monetary Theories	
Free markets	Managed currency exchanges
Monetarism	Low interest rates
Free trade	Growth in money supply
What Validates Money	
Shortage of currency	Excellent work
Basic Economic Theory	
Market determines value	Design determines value
Wealth is gathered	Wealth is manufactured
Goal of Economics	
Wealth is to be concentrated	Wealth should be widely spread
Truth Tellers	
Constitutional scholars	Scientists
Theologians	Mathematicians
Gurus	Engineers
Mystics	Builders
Ideologues	Inventors

PREDATORS	PRODUCERS
Means to Truth	
Scholarly examination of previously defined "truths"	Research
	Experimentation
	Dedication
Discipline	Curiosity
Obedience	
Schools	
Harvard University	Massachusetts Institute of Technology
Oxford University	
Their imitators	M.I.T.'s imitators
Get rich in real estate courses	Vocational Schools
Definition of Intelligence	
Cunning	Creativity
Validation of Knowledge	
Footnotes	Experiments
Appeals to authority	Patents
Information Exchanges	
Propaganda	Instruction manuals
Public relations	Video cassettes
Mass media	Photocopy networks
Scares	Computer bulletin boards
Philosophies	
Idealism	Materialism
Dogmatism	Pragmatism
Determinism	Free will
Futurism	
Read "Revelations"	Invent
Indulge in games of chance	Plan
Play the stock market	Build
Frame of Reference	
Extremely short	Very Long
History	
Thugs and thieves	Everything else
Downfall	
War	Environmental destruction

Is Conflict Between Producers and Predators Inevitable?

Some people find themselves to be a natural mixture of the value currents of both classes. When predator and producer values are mixed, either within a person or group, there are three possible outcomes; two of them are bad. Leisure and industrial values can collide which will cause destruction or stalemate. A forced compromise between leisure and industrial values can cause mediocrity.

Because the values of the leisure class are so visible, they are often seen in industrial design. Product planning of the 1950s, which brought the United States oversized auto-boats with tailfins is a perfect example of leisure-industrial value mixtures leading to mediocrity.

Leisure class values often creep into industrial enterprises as they mature. This is rarely a cooperative venture. The leisure class values usually arrive with the predators who seek to harvest the fruits of the industrial enterprise. The tensions caused by such predatory practices are usually enough to render such an industrial enterprise permanently uncompetitive in an industrial sense. From that point on, it is usually a process of ever more violent plunder until the last of the great predators, the liquidators, come to pick the bones in the great example of stalemate and then destruction.

It is the third possible outcome that is most interesting. There are occasions, such as the building of a great concert hall, when leisure and industrial values are added in such a way that the outcome far exceeds the expectations of either group. Most unfortunately, these occasions are extremely rare because when they happen, everyone is a winner. The world is in shocking need of many more win-win situations.

It is in the area of environmental concerns where the values of the hunters and farmers must come together. It is the last great hope for cooperation.

So far, there has not been a great amount of cooperation between the producers and the predators when it has come to environmental concerns. The producers have dug in their heels and told the world that they are essential and that the world will just have to learn to live with the mess they make. The predators have told the producers that lack of cooperation means a shutdown of industry. Pittsburgh was recently named the most livable city in the United States by Rand McNally. Pittsburgh was once an extremely dirty city. The difference is that the great steel mills of the Monagahela Valley have been closed down and left to rust—not a very good solution from the standpoint of the steelmakers, but effective.

Shutting down industry is not the solution. Whether people believe the industrial revolution was a good idea or not; it happened. With the industrial revolution came an enormous population rise. If industry is indiscriminately closed, the dense populations that are the result of industrialization are in great peril of their very survival.

A better way to view the industrial revolution is to see the current industrial class crises as the *halfway point*. If the world turns back to an early stage of industrialization, there will be a massive reduction in the human population. If the present confusion of industrialization and its values is not clarified, there will be also a massive mortality of humans.

The solution is clear, the world must build its way out of the current problems using an environmental blueprint. Industrial values must be environmentally purified. Hope stems from the realization that this value purification process, while still in its infancy, has already appeared in the sophisticated industrial economies of northern Europe.

VALUE CONFUSION VERSUS VALUE CLARITY

ARTIST. GENIUS. CREATIVE. GODLIKE. Producers have been treated poorly throughout history, but they have developed a fine lingo to describe themselves.

The downside of industrialism is not so pretty. Producers must answer grave questions. Whose "bright" idea was it to produce fission nuclear power plants? or agent orange? or strip mining? or clear-cut timbering methods? or nerve gas? or? or? None but the producers can answer these questions.

The producers' response is to counterattack. "These were my ideas," argues a producer, "but they were not my decisions. There were many other ideas. Those monuments to industrial stupidity exist because they were funded. A producer can build anything. What gets built is decided by people with money."

"Whose bright idea?" asks the producer, "was it to insist on economic theories that pretend to describe a system built by industrial planning, yet claim that industrial economic planning is impossible or evil? Who possibly believed that flying, the human activity most in need of careful regulation, could be deregulated economically? What fool could fall for the notion that a nation needs more than one telephone system, which mega-fool thought Bell Labs was unimportant? Didn't anyone understand that if one country funds 15-year projects and another funds 1-year projects, the 15-year products will be far superior? Why should anyone believe that Wall Street plunder of the needed resources for research and development would not lead to industrial stagnation? What took you predators so long to understand the Cold War was over? It was over technologically by 1960. It is not our fault that everything had to be labeled a 'war' on something or other or it would not be built. If you want a utopia, we can build a utopia, but if you want to have us solve those problems you so willingly lay at our feet, the first order for business is to get off our throats and let us go back to work. To us involuntary unemployment is the ultimate human-rights abuse!"

The implication is that if environmental redesign spawned the well-funded, high-status new producer professions, producer elites would retrain themselves to meet the need. This redistribution of industrial talent to environmental design is the obvious answer to the problem of civilian conversion of militarized industries. Only producers can solve the great environment problems: only they have the requisite skills. Until the economic rules are changed, however, they cannot do their work.

If the producers were granted a set of economic rules that assumed the industrial revolution, the question still must be asked, "Would producers, given their freedom, build an operating industrial-environmental state?" The masters do not trust the slaves. Like the homeowner supervising the remodeling of a kitchen, predators are not certain the producers will do the work properly.

As the kitchen example shows, the friction between producers and predators can be finessed. Good architecture requires a good client. The finest example of producer-predator friction yielding great results may be Michelangelo's painting of the Sistine Chapel. A prototypical producer, Michelangelo *despised* Pope Julius II and insisted on doing the painting his own unique way. The Pope did his part—he paid the bills, he forced a creative person to express himself in a new medium, and for the most part, he stayed out of the way.

Not all producers are Michelangelos unfortunately; however, the 1980s demonstrated that there are far more Michelangelos than enlightened predators. There will be no environmental Michelangelos until the producers are free to do their work. The job of the environmentally concerned predator is to demand excellence but to stop wanting everything on the cheap, pay the bills, and stay out of the way. It is a strategy that has worked in the past.

Making the producer-predator relationship more enlightened is a necessary strategy in those countries, such as the United States, which would play industrial catch-up with Japan and Germany. Killing the predators in war—the "solution" of industrializing Japan and Germany—might appeal to producer *schadenfreude*, but is not a desirable method in a world with 50,000 extant nuclear weapons. Besides, producer creativity itself is highly vulnerable to war.

A better strategy would be a widespread purification and socialization of industrial virtue. In economic terms this means that the industrial-environmental projects necessary for human survival must be funded and produced. Economics must be redefined so that what is necessary is economically valid. Economics must be changed so that finance and business are the servant of, rather than an impediment to, the construction of the new industrial order.

American industry has acquired a bad habit in the twentieth century by being forced to sell large projects in military terms. From scientific education and the interstate highway system to the space program, militaristic salesmanship has been employed. Until producers learn to sell large projects without resorting to a militarized sales strategy, predators will never learn to evaluate projects in any other way. Rocket scientists stampeding dull-witted politicians for financial support may be an amusing spectacle, but until producers learn to market large social projects for their intrinsic merits, they will always be slaves in the societies they have built for themselves.

NOTES ON CHAPTER FOUR

[1]Jacob Bronowski, *The Ascent of Man*, Little, Brown and Co. (1975), pg. 116.

This quotation is, in many ways, the theme of this book. If Bronowski is correct in his belief that humans will evolve as they teach themselves to use tools, then a tool-driven culture, such as would please Veblen, is very likely. For those who cannot find Bronowski's book, here is the whole quotation.

> We are active; and indeed we know, as something more than a symbolic accident in the evolution of man, that it is the hand that drives the subsequent evolution of the brain. We find tools today made by man before he became man. Benjamin Franklin in 1778 called man a 'tool-making animal', and that is right.
>
> I have described the hand when it uses a tool as an instrument of discovery; it is the theme of this essay. We see it every time a child learns to couple hand and tool together—to lace its shoes, to thread a needle, to fly a kite or play a penny whistle. With the practical action there goes another, namely finding pleasure in the action for its own sake—in the skill that one perfects, and perfects by being pleased with it. This at the bottom is responsible for every work of art, and science too: our poetic delight in what human beings do because they can do it. The most exciting thing about that is that the poetic use in the end has truly profound results.
>
> ...The hand is the cutting edge of the mind. Civilization is not a collection of finished artifacts, it is the elaboration of processes. In the end, the march of man is the refinement of the hand in action.
>
> The most powerful drive in the ascent of man is his pleasure in his own skill. He loves to do what he does well and, having done it well, he loves to do it better. You see it in his science. You see it in the magnificence with which he carves and builds, the loving care, the gaiety, the effrontery. The monuments are supposed to commemorate kings and religions, heros, dogmas, but in the end the man they commemorate is the builder.

[2] Tom Wolfe, *The Worship of Art*, Harpers Magazine (October, 1984). *From Bahaus to our House* Farrer, Straus, Giroux (1981).

Tom Wolfe deserves a great honor for writing these two pieces. In *The Worship of Art*, Wolfe devastates the art-of-the-latest-theory. He prefers the art-of-the-high-craftsman, and so does most of the rest of humanity. More subtlely, he suggests a huge fraud. A company which plunders the earth, provides dismal working conditions for its employees, and constantly lobbies for lower taxes so the city is a blighted mess—all in the name of profits, and then spends some of those profits for a tangled mass of misshapen steel, and does so in the name of

NOTES ON CHAPTER FOUR

beauty, is really fooling no one. Wolfe gets through the haze of modern art like no other. Moreover, his is about as enjoyable as reading ever gets.

From Bahaus to Our House borders on being, if it is not, a very important work. His critique of modern architecture is devastating. Architecture is incredibly important. To allow egomaniacs to design environmental disasters in the name of their good art is a mistake of monumental proportions. Wolfe blows the facade off the edifice of modern architecture. If architecture can be redirected towards more useful goals because of this work, Wolfe may rank as one of the heroes of the twentieth century.

[3] There was a time, not so long ago—before Shriners as clowns and Propaganda Due, (P2) when the Freemasons were the soul of the Enlightenment. George Washington, Thomas Jefferson, Voltaire, as well as Mozart and Beaumarchais were all Enlightenment Freemasons.

Because they were a secret society and hostile to the religious establishment, Freemasons have gained a conspiratorial image—a image modern Freemasons do not easily dispel. When they were the guiding force of the Enlightenment, however, they were as socially progressive as any movement in history. Jacob Bronowski records this Beaumarchais sentiment in *The Ascent of Man*, pg. 268 (Beaumarchais wrote *The Marriage of Figaro*—often called the play that spawned the French Revolution)

> Because you are a great lord, you think that you are a great genius. Nobility, wealth, honors, emoluments! They all make a man so proud! What have you done to earn so many advantages? You took the trouble to be born. Nothing more. Apart from that, you're a rather common type.
>
> A public debate started on the nature of wealth, and since one needn't own something in order to argue about it, being in fact penniless, I wrote on the value of money and interest. Immediately, I found myself looking at...the drawbridge of a prison...Printed nonsense is dangerous only in countries where its free circulation is hampered; without the right to criticize, praise and approval are worthless.
>
> Beaumarchais represented a different aristocracy, of working talent: the watchmakers of his age, the masons in the past, the printers. What excited Mozart about the play? The revolutionary ardor, which to him was represented by the movement of Freemasons to which he belonged, and which he glorified in [*Figaro* and] *The Magic Flute*.

[4]Joseph Borkin *The Crime and Punishment of I.G. Farben*. Free Press (1978)

The case of Karl Bosch is an interesting example of an industrial supermind being a political idiot. Chemistry is Germany's strongest industrial suit. The various chemical industries combined their energies in a conglomerate called I.G. Farben. Karl Bosch helped invent a process called the Häber-Bosch which would allow for the synthesis of nitrate fertilizer. At the time this process was

NOTES ON CHAPTER FOUR

invented, Germany was not interested because they were getting all the nitrates they needed from their colonies in Chile. When the Germans entered World War I, they assumed that it would be a quick war and had provided only enough explosives for six weeks. Nitrates for fertilizer and nitrates for explosives are the same thing and the British Navy, operating out of the Falkland Islands moved to shut off the supply from Chile. Enter Bosch. In less than a year, he was able to take a successful laboratory experiment and convert it into an industry large enough to keep the war going at murderous rates until 1918. This was a phenomenal accomplishment. The distance from the laboratory to full-scale production is usually measured in decades. For this accomplishment, Bosch was elevated to the board of directors of I.G. Farben.

As Hitler prepared for World War II, the German chemical industries figured to play a major role. They were critical. The *Luftwaffe* and *Wehrmacht* would roll on tires made of synthetic rubber, powered by synthetic gasoline, and kept going with synthetic lubricants. Bosch viewed these assignments from a technical perspective and tried to ignore the politics. In his defense, the ordered mindset necessary to accomplish what Bosch had done made the madness of Hitler utterly incomprehensible. He heard what he wanted to hear. When Hitler talked about the "Jewish Problem," he thought of the effect of the currency collapse on industry, not the Jewish chemists who worked for him. An epic story.

Selective hearing is still a major problem of industrial interests. American farmers voted for an administration in 1984 which fostered policies that would bankrupt most of them. When Reagan attacked the Welfare State for breeding social parasites, they all cheered. They weren't parasites, they were producers—and damn hard-workin' ones at that! Many used to be millionaires and Reagan talked like a proud millionaire. They thought he was talking like a producer.

5 William Manchester, *The Arms of Krupp,* Little, Brown and Co. (1964).

The House of Krupp dates from the sixteenth century. They made steel. In 1870, heavy guns supplied the Prussians by Krupp effectively blew up the French army thereby ending the Franco-Prussian War. Krupp made, and makes, extremely fine *Kruppstahl*. His guns were better the the General Staff could understand. Hitler understood their frustrations and promised to let them build the best guns they could. (In the end, this was part of the undoing of the German army. Give industrial types their head and they can get fancy. Simple Russian guns beat complex German guns in the mud of Russia.)

Happy with this arrangement, Krupp interests backed Hitler. They were rewarded with the spoils of war. They were provided with forced labor and built a plant at Auschwitz. The heavy irony of the Krupp experiment with Fascism was that it did not pay. The Krupps feared the Social Democrats and to keep them out of their facilities, they used repression and a compensation package that many German workers found quite acceptable. If you worked for Krupp, your children were born in a Krupp hospital, educated in a Krupp school, played on a

NOTES ON CHAPTER FOUR

Krupp playground, etc. You shopped in a Krupp store. It was cradle-to-grave security of sorts. Women advertising for husbands would specify employees of Krupp. During World War II, these "expensive" employees provided the Krupp works with steady profits. The forced labor of Aüschwitz always posted a loss.

Leaving aside the lessons of Nazism, the Krupp industrial model is interesting. It might be called "industrial paternalism." By provided the services necessary for life to employees, the industry can grow irrespective of the money games that swirl around it. Since industrial growth has often been hampered by shortages of currency, this is a useful tactic. The problems with industrial paternalism is that not everyone likes this arrangement and further, it is very sensitive to the quality of the top manager. Even so, there are American examples. Phillips Petroleum, arguably the most innovative in the business, was for years, a shining example of industrial paternalism with many essentially contented employees. There are many ways to organize industry.

The Krupp industrial doctrines were summed up by Manchester as follows: pp. 146-47 (Interior quotes from the *Generalregulativ* itself.)

> For nearly a century the *Generalregulativ* was to remain the Konzern's basic constitution. It is not too much to call it a blueprint for all of German industry. Everything that was to emerge in the decades ahead—the rigid chain-of-command system, vertical and horizontal integration, the establishment of cartels—was tersely set forth in its fine Gothic script.
>
> The full force of authority must be used to suppress disloyalty and conspiracy. Those who commit unworthy acts must never be permitted to feel safe, must never escape public disgrace. Good, like wickedness, should be examined through a microscope, for there truth is to be found.
>
> Employees are expected to render "full and undivided energy, loyalty, a love of good order, and a freedom from "all prejudicial influences." This last included a "refusal to work or inciting others thereto."
>
> What may strike the *Ausländer* as odd is that Alfred's (Krupp) General Regulations were regarded—and in Essen are *still* regarded—as liberal. For the first time a German firm was spelling out it duties to its men. Kruppianer could lay claim to "a health service, a relief fund...a pension scheme, hospitals, a homes for the aged" and though this would not come into effect until 1877, Krupp's "Life Insurance Institution." Nothing remotely resembling this may be found in the archives of the other titans who were emerging from the industrial revolution.
>
> Low cost housing, a bread factory, a wine store, a butcher plant, a hotel, and a charity fund for families left destitute by the flooding of the Ruhr, soup kitchens and public works were provided for the unemployed during the Great Depression. Konsum-Anstalt—a non-profit chain of cooperative stores served all Krupp employees.

NOTES ON CHAPTER FOUR

> ...Such paternalism, Norman J. G. Pounds pointed out in his study of the Ruhr, "was contrary to the social and political development of the time." Krupp meant it to be. ...Alfred sent the Kaiser a copy, which survives in the Krupp family archives; on the title page, in his bold, jagged hand, is the inscription, Originally determined for the protection and flowering of the works. Besides that, it is useful for the prevention of socialistic errors. Among those who saw the moral was Wilhelm's chancellor. The parallels between Alfred's text and Bismarck's social welfare legislation of 1883, 1884, and 1889 are unmistakable.

6 *Business Week* is a publication with many industrial class values. There are those who may wish to contest the point but the issue of September 16, 1985, with its cover story "The Casino Society," clearly put *Business Week* in the camp of those who abhor the money games intruding on the real workings of the economy. That is "industrial-class" enough for this book. Of all traditional business publications, only *Business Week* lifted its voice in protest over the plunder of industry in the 1980s.

PART TWO:

THE ECONOMICS OF INDUSTRIALISM

CHAPTER FIVE
THE ECONOMICS OF PREDATION

BASHING ECONOMISTS has become one of the most popular indoor sports. This sport goes on in most nations on the planet. The cause is real. Most of the world finds itself in dire economic conditions. The same economists who were so eager to take the credit for the sunny days of industrial expansion following World War II, have found their most basic premises under attack as the boom turned sour.

Most Americans who took economics in college during the 1960s used a textbook written by a Nobel Memorial Prize [1] winning economist named Paul Samuelson—in fact he won the prize in part for the textbook. In his 1961 edition he wrote that the body of neoclassical theory is "accepted in its broad outlines by all but a few extreme left-wing and right-wing writers." This is confidence that borders on arrogance. The central currents of economic thought held such sway that someone could make such remarks and be taken seriously. The economists believed that their understanding was so profound, that the situation was under control, and all that was necessary for them to do was fine-tune the economy. In the context of the times, such hubris was understandable. Most economies were performing better than at any time in history. It should be noted that the health of the economy probably had little to do with the prescriptions of economists.

During this period, Daniel Bell, an historian at Harvard, found himself so caught up in the spirit of the times that he announced that the problems of production had been solved. The industrial democracies had demonstrated their overwhelming technological prowess to such an extent that the time had come to quit the battle and march forward bravely into the post-industrial age where the most important skill would be information processing. The concept of post-industrialization soon gained a degree of favor that far exceeded its worth as an idea. Post-industrialization assumes that because the number of persons employed in industrial occupations is shrinking, their activities are increasingly unimportant. Using straight-line trend projections, the post-industrialists seem to think that at some point, the number of industrial activities will reach zero. As is obvious, this assumption is flawed to the point of absurdity. Because of this, the concept of post-industrialization has been so thoroughly diluted so as to mean almost anything or nothing at all.

As the economies of the world fell apart in the 1970s, the consensus in economic thought became fragmented. Now, not only were fringe elements found outside the body of neoclassical thinking, but economics as a profession found itself split into neo-Austrians, neo-institutionalists, neo-Marxists, Chicago monetarists, and post-Keynesians along with a host of backward-looking fundamentalists. This last group does not even feign any pretense at objectivity but is only in the business of rationalizing privilege.

Faltering economies promote unstable politics and in country after country, voters—at least in those countries with voters—chose to try parties with alternate programs. The French voted in a Socialist government that turned out to be Keynesian. The British replaced the Labor government with Thatcherism, a form of Chicago monetarism. The examples are too numerous to cite. All peoples with the opportunity have tried to change their government on the promise that a new government comes with a new economic game plan. The result has been a period of economic experimentation almost without precedent. Virtually every form of capitalist economics has been tried somewhere in the world. Strangely, the most successful capitalist economies of Japan and Germany are not often imitated and their economists do not win Nobel prizes.

The Marxist economists had an even more interesting conundrum. While gleefully pointing out the problems of capitalism, the problems of capitalism were their own. One of the articles of faith among Marxists was that socialism is as superior to capitalism as capitalism was to feudalism. According to Marx, the only reason for capitalism to exist in a historical sense was that capitalism was the stage where the problems of production are solved. If a capitalist system breaks down because of problems in the systems of production, especially if the problem is a technological mismatch caused by the shifting realities of resource supply, for example, then any problem that appears in capitalism will eventually appear in the Marxist states.

Although the Marxist states were known for many things, technological innovation was not one of them. Not only did the

Marxist states fail to come up with an alternate industrial model for themselves, they were unable to execute the western model. The closest comparison would have been between East and West Germany. When the wall came down, the East Germans were making an ugly little car called the Wartburg, while the West Germans were making awe-inspiring vehicles like the BMW 750il.

In the main body of Marxist thought, the problems of industrialization are dealt with only as a problem of social organization, an enormous contribution to the understanding of industrial problems, to be sure, but hardly the whole story. The technological and scientific problems of industrialization were beyond the expertise of Marx. He was neither an engineer or a natural scientist, so he put them in a kind of limbo where they languished, much to the detriment of industrialization in the Marxist states.

Of course, there are those who will claim—most notably the Trotskyites—that Marxism was never really tried anywhere in the world and if it ever were to be, it would prove itself superior—especially if had grown out of an advanced capitalist state as Marx had intended. Taking all the governments that claimed to be Marxist and assume they were what they claimed to be, it would be fair to say that in many ways—especially technological, their economies had the same problems as those of the capitalist democracies.

The stunning reality of all this is that virtually all economists, Marxist and capitalist alike, lost their compasses simultaneously. Productivity growth is a problem in the former Soviet states. They blame corruption, obstructionist bureaucrats, or lately, lack of market incentives. Productivity growth is a problem in the United States The common explanation is lack of investment.

Just at a time when the old pet theories of how the world works are failing economists in both east and west, comes, of course, the calls for solutions. Economists the world over must be wishing they had not taken credit for the boom of the post-World War II period. Technically, of course, economists are supposed to be scientists. As scientists, their job is to describe the currents and motions of the economic activity around them much as an astronomer charts the motion of heavenly bodies. Compared to the astronomers, the job of the economists is infinitely more difficult. Contrary to the most fond hopes of the most dedicated supporters of econometrics, human behavior is much too arbitrary to be modeled.

Like the discovery of retrograde motion in the planets that led to the doubts about the model of the earth as the center of the universe, so the shocks of the 1970s and 1980s have been to the economic models of industrialization. The distress shows in the titles of current economic literature. *What's Wrong With Economics, Economists at Bay, Why Economics is Not Yet A Science,* and *Dangerous Currents: The State of Economics* are all recent titles. It is hard to predict anything if the models fail to define the current realities.

Leaving aside the forecasting function, the study of economics is really a form of historical pulse-taking. The data examined by economists are, at best, a day old. Whatever an economist describes is something that has happened in the past. What this means is that millions of critical economic decisions have been made by quite ordinary citizens before the data gathering even begins. Not only are economists having a great deal of difficulty with the prescriptive function of their job, they are having problems with the descriptive element. [2]

THE MYTH OF THE MARKET

BELIEF IN THE NOTION of a free market is the distinguishing characteristic of the leisure-class capitalist economist. More than that, in the United States, belief in the free market is the boundary definition of the economics profession. Those who doubt this basic thesis are cast out of the profession and are labeled economic philosophers, anthropologists, satirists, moral philosophers, or anything but economist.

The notion of the free market is elevated to the status of untouchable scientific law because with it one can mathematically model human behavior. This ability to model human behavior is why economists believe themselves to be more scientific than their fellow social scientists. If the laws of supply and demand are not true, then economists fall back to the level of other mere social scientists.

When the bright-eyed student arrives for the first economics class, the first item on the agenda is the parable of the free market. The story goes something like this:

There is a fisher who goes to market with his catch. In order for him to sell all his fish, he must set a price that is low enough to clear out his fish stand by the end of the day. If he sets the price for the fish too low, he will not maximize his return. If he sets it too high, he will have fish left over at the end of the day which will spoil and cause losses in that manner. To be successful in the marketplace, the fisher must get the highest possible price while still selling his entire catch. Of course, if other fishers had good catches and the market is flooded with fish, the price must fall because everyone will be trying to clear out supplies. If the fisher was the only one lucky enough on a bad day to have any fish to sell, he can raise his price to whatever level he chooses.

This parable has a certain amount of believability. If the economics student goes down to the local fish market for several weeks in a row, he or she might see just this sort of supply-based price adjustment taking place. With this confirmation, the student can return to the classroom secure in the knowledge that the market controls the price. If the level of supplies influences price and price influences demand, then there is a relationship between supply and demand that is predictable.

If the economics student believes this parable of the free market law of supply and demand, the first big hurdle has been jumped on the way to becoming an economist. If the student happens to make the mistake of going down to the supermarket instead of the fish market, the likelihood that the parable of the free market will be confirmed diminishes rapidly. Certain products, like fresh blueberries, will conform to the laws of supply and demand while others, like milk and cheese, seem exempt. Doubts will form because not all products seem to react to the same law. The student returns to class with these doubts and is assured that, yes, milk prices are an exception that will be covered in a later chapter, but before an understanding of something as arcane as milk price supports can be achieved, the basics must be learned. The student must be able to walk before running. Milk prices are the exception that proves the rule, but the laws of supply and demand work, the student is assured.

The student could make the mistake of getting to the right fish market but talking to the wrong people and stumble upon the fact that setting the correct price is not the only element in selling all the catch. The wrong people to talk to would be the fishers. If some fisher took the student into his confidence and told his tricks of the trade, the student might come up with questions his professor could not answer. If the student were to think about the market as a fisher instead of a budding economist, he or she might ask "Suppose my fish were not selling well enough to clear my fish stand by the end of the day and I didn't want to lower my prices, would I have any alternatives?" Given the amount of exposure to sophisticated marketing techniques endured by the average economics student in any industrial society, the student could well come up with many alternatives gleaned from personal experience. Come up with enough plausible alternatives and he or she will be doomed as an economics student:

1) The student-fisherman might decide that the price he or she set for the fish was fine, but the sales location was bad. One alternative would be to find a friendlier spot in the fish market.

2) It might be decided that the fish stand looked a little ratty and needed to be scrubbed down and painted.

3) The sign could be enlarged.

4) A child could be employed as a barker.

5) The fisher could get fancy, rename his catch and pass out exotic sounding recipes. This, of course, is outside the economic rules because the customer is supposed to know enough about buying fish not to be fooled by such ploys. The historical fact is that when the North Atlantic cod fishers were suffering from an inability to market their catch, they renamed their cod (which is, after all, a very unappetizing sounding name for food) "torsk" which sounds vaguely foreign and is another but unfamiliar name for cod. They printed recipes that touted torsk as the poor man's lobster and placed them next to the fish cases in the supermarkets. This marketing scheme was so effective that 15 years later in some parts of the United States, though the various consumer agencies have forced the label of cod back onto the frozen fillets, people still ask for torsk in the stores. Recipes in the newspaper food sections will include the term "torsk" along with a disclaimer that torsk is really North Atlantic cod. More interestingly, this was the first major inroad in the Midwestern American beef diet. Fish was on its way to becoming a major dietary player. This example amply demonstrates that in many cases, markets are manufactured because demand is manufactured. [3]

6) The fisher could decide that since business was slow, he could spend his time filleting the fish: so that instead of lowering the price of his fish, he would be raising the value. Of course, this is outside the supply and demand parable because when the fisher fillets his catch, he is no longer in the whole fish business but the processed fish business. But modifying what has been gathered from nature is a most common feature of human endeavor. Placing such endeavors outside the parameters of the basic parable weakens its impact considerably.

7) The fisher might decide not to lower his price and give away what is left of his catch to the orphanage on his way home. He might reason that lowering the price can get to be a bad precedent. If the word gets out that the fisher always lowers the price at noon, folks will refuse to buy until he runs his clearance sale. The price must be kept up, even if some of the catch goes home to the compost heap because tomorrow's market prices must be considered. The only description for such behavior by economists' thinking would be "irrational."

A realization begins to gnaw at the student-economist that the only form of human behavior allowed under the laws of supply and demand, the lowering of prices, is the *last* option the real-world fisher will try. It is in the interest of the real-world fisher to do everything in his power to invalidate the very laws that are supposed to be so universal that mathematical models can be constructed around them. It occurs to the student-economist that if the exemplary market of the parable can be manipulated, manufactured, altered, and circumvented in the real world by people who have real-world motivations for doing so, then imagine what happens in the markets for more complex products.

The parable of the fish market is so appealing to economists because it is so simple. Buyers and sellers in such a market have roughly the same amount of information and power. There is perfect competition and perfect knowledge. Renaming cod "torsk" may help get people to try a fish they might not have otherwise tried, but once they got into the habit of eating "torsk," the laws of supply and demand would reassert themselves as folks made their buying decisions based on price. No matter what the fish is called, eventually the decision comes down to easily evaluated variables such as size, freshness, color, smell, and so on. Those who have the intelligence to feed themselves should be able to buy fish with as much skill as anyone else. It is possible, in economic terms, to have perfect knowledge of fish buying. It is also possible, in economic terms, to have a form of perfect competition if there are enough buyers and sellers.

Fishing is a gathering occupation dating back before recorded history. For all practical purposes, people have always fished. Fish, with a few modern exceptions, are what they are without any intervention from humans. Humans can alter the shape of fish by cross-breeding but fish are not something that humans create. Humans create automobiles. Humans cannot create fish. Because the forces involved with the creation of fish are superior to the forces mustered to create an automobile, the variations within species are so obvious and few, that virtually any conscious human with any interest can master them. For reasons of setting price, perfect knowledge is possible.

Perfect knowledge is clearly not possible in the case of products made by humans. Comparing cod A with cod B is child's play compared with comparing automobile A with automobile B. This is so because those who could have anything to do with making comparisons difficult have done *everything* in their power toward reaching the goal of making sure their products can only be compared with themselves. That is the *whole point* of modern industrial enterprise. Economists are by definition of the hunter-leisure class when they assert that the gathering-distribution parable of the free market has everything to do with the industrial economies of the twentieth century.

A person who tries to apply the lessons of the fish market to the problem of buying a car is confronted by realization that perfect knowledge is not merely difficult, but impossible, to obtain. If a leisure-class economist goes to market to buy an automobile with the clear goal of keeping transportation cost per mile as low as possible—maximizing utility, he will find the problem of comparison so difficult that a mathematical model of the comparison process probably will get him published in a scholarly economics journal.

The person wanting to compare the cost of owning car A with car B must first come up with a working definition of the useful life of the purchase. One cannot evaluate the car as a whole but only as a collection of individual components. The car's useful life is ended when the buyer sells the car or when it is junked. If the buyer intends to sell the car at some future date, the cost-per-mile calculations must include projections of potential selling prices so as to account for depreciation. This is a form of forecasting that can, at best, be a version of trend projections based on historical data. A trip to the library can give the car buyer some idea of past depreciation rates of various automobiles. There are complications to these data. If the buyer is thinking about a new model, past depreciation rates cannot be relied upon. If the buyer is thinking about an existing model with an established track record, the risk is that a new and superior model will be introduced before he sells and his purchase will have fallen from favor. Although the current depreciation rate may be true, it will be dramatically wrong in the future.

If the decision is to keep this car until the wheels fall off, the buyer must consider the future maintenance costs. Such calculations would include price of spare parts, frequency of parts replacement, costs of labor, and the real life expectancy of irreplaceable parts. The buyer must understand that replacement parts will not always be available because either the carmaker will go out of business or stay in business and halt spare parts production. Because of the difficulties in obtaining spare parts, keeping a 1932 Ford or a 1953 Studebaker running is an expensive hobby, not a low cost-per-mile form of transportation. Armed with these educated guesses, the buyer must be able to predict when the car will cost more to maintain than to replace to arrive at some form of useful life calculations.

Having made an educated stab at depreciation and maintenance costs, the economist-buyer can turn to the easier parts of the equation like comparing fuel efficiency with purchase price. If fuel prices hold steady, a simple calculation can be made of how long it will take for an increased purchase price plus finance costs to be returned in fuel cost savings. Of course, since petroleum cannot be created but only gathered and refined, it does conform to *some* of the laws of supply and demand and can be projected to fluctuate. The one law the economist-buyer sincerely believes in now threatens the easiest part of the calculations.

Then there are the esoteric calculations. What are child-proof rear door locks worth if they keep a child from falling out of a moving car? What are good seats worth if they keep the driver from getting a bad back? What are good brakes worth if they keep the car out of a fender-bender? The future economic well-being of the economist-buyer may well hinge on questions he is not in any position to ask much less answer. The notion of perfect knowledge, in an industrial setting, is patently absurd. Since perfect knowledge is a key underlying assumption of the free market parable, the notion of the free market in an industrial setting is equally absurd.

Perfect knowledge and perfect competition are not the only forms of perfection in which leisure class economists believe. Some adhere to the notion of a perfect labor market. The perfect labor market implies that all labor is the same and can be easily inserted or deleted wherever the need arises. Of course, the notion that an unemployed supermarket worker in New Jersey can become a systems computer designer in California without a huge investment of money, time, and effort means the perfect labor market idea is shot full of holes. One would think that the very word perfect would be enough to make the leisure class economist nervous, perfection being a rare commodity. Industrial minds believe in manufacturing tolerances, not perfection. They look on arguments with the word "perfection" in them as excuses to close the debate early. Who, after all, can argue with something that is perfect?

The charge that economists use words like "science" and "perfect" as smoke screens to hide much more questionable underlying assumptions is valid. Not all the smoke screens are deliberate efforts to deceive, however. Most are forms of confusion caused by bad reasoning, worse premises, word-traps, irrelevant analogies, and lack of information. Confusion or deception; it matters not.

THE ECONOMICS OF PREDATION

THE ULTIMATE FAILURE of leisure-class economics, however, is that it is a static system. The marketplace is a wonderful place to take an economic snapshot of current conditions, but as Lester Thurow points out so magnificently in "Dangerous Currents," *there is no market between the present and the future.* [4] This is sufficient reason to reduce the notion of the market to its role in determining blueberry prices. The only valid statement that can be made about the future and the market is that the market will set the price of blueberries in the future. Nothing more.

What is simply a minor oversight in an economic description of the market as a distribution system for gathered wealth, becomes a fatal flaw when the same economics is applied to industrialization. In the gathering-distribution economic market description, the creator is overlooked. How iron ore was formed is the province of the geologists or possibly the theologians, not economists. How it was formed is unimportant in the economic scheme of things compared to who owns it. Forgetting the creator may be a harmless mistake with iron ore, troubling mostly sentimental clergymen. *Ignoring the creators of manufactured wealth and how they create is to ignore most real economic activity in industrial states.*

The world of the economic creator is not a static world. By its very nature, creativity is a dynamic activity. Static market considerations applied to creative industrial activity are simply inadequate. Moreover, creative industrial activity involves planning—the only bridge between the present and the future. As a result, market economics and planning are mutually exclusive.

The result is quite amusing. Huge economic textbooks, filled with laws, axioms, postulates and other wise sayings from the static, simplistic view of the world, devote only a few pages to the creative, dynamic side of the economy. Such attention is usually confined to an observation about research and development budgets—how appropriate and how much. Economists labor long and hard over the human behavior involved with dividing the pie but ignore the human efforts to produce the pie.

THE MYTH OF VALUE-FREE SCIENCE

IF ECONOMICS IS NOT THE SCIENCE it pretends to be, it must be something else. From the industrial point of view, economics looks like a form of religion complete with articles of faith. The articles of faith of the hunter-leisure class make perfect sense in a predatory setting but cannot stand up in a modern industrial setting and must be examined.

Article of faith #1: It is possible to be rich when your neighbor is poor.

"I'M ALL RIGHT, JACK," a British expression that covers this sentiment perfectly, conveys the hunter-leisure class feeling that survival is difficult enough without worrying about anyone else. This feeling is natural enough. For millions of years, people survived by looking out for themselves and their families before worrying about the problems of the rest of the clan. The great hunters prospered. The weak fell by the wayside. This is social Darwinism taken to its logical conclusion.

In industrial economies, the problem is not so simple. There may exist a difference between rich and poor—such differences seem inevitable, but if the differences between the rich and poor become too great, the mechanisms for becoming and staying rich disintegrate. In an industrial setting, the poor cannot become too poor because they lose their usefulness as consumers and the industrial wheel slows down. Historically this has been true. Panics, recessions, depressions, and other forms of industrial economic upheaval have followed great separations between the wealthy and the poor. On the other side of the coin, the societies with the smallest multiples between the best and worst paid members of society, like Japan, are the strongest industrially.

There is an aesthetic rejection to the possibility of being rich in a sea of poverty. How rich can a person be who must drive through dirty, ugly slums, teeming with people who would kill him or kidnap his children? There must be more pleasurable life styles than living behind walls, bouncing around in an armored car, and sending the children to foreign countries for an education. This is how the rich live in Central America and South Africa. The whole point of colonialism and absentee landlordism is to avoid this problem. Move with the money to Switzerland or Palm Springs and leave the ugliness of poverty behind. Hire a bunch of predators to hunt down and kill anyone who disagrees with this arrangement and it is possible, on a *temporary* basis, to be rich while your neighbor is poor. While excessive gaps between rich and poor merely push industrial societies into recessions, huge income gaps in the states of neocolonial feudalism cause revolutions.

It should be noted that most real indicators of wealth are *public*. Having an expensive car is of very little use unless there are good roads to drive it on. This obvious point is lost on the predators who believe that wealth is strictly a private measure.

Article of Faith #2: Anything is permissible if it leads to your personal survival.

WHEN THE FIGHT was between humans, armed with rocks and sticks, and large and very dangerous animals, anything the human could think of was permissible. There were no ethics when cornered by a bear. Correct that. There was one ethic: get out of the corner alive and if there are bear steaks for dinner, so much the better.

The problem is that industrialization has enlarged the definition of anything. From thumbscrews to the rack, from Aüschwitz to the atomic bomb, humans have become increasingly barbaric as their possibilities for barbarism increase. A perfectly useful survival instinct of the hunter class can lead to perfectly awful behavior.

Personal survival as an economic rationalization causes serious problems and even more serious failures of understanding. The real reason economics is called value-free is because economists refuse to make any judgements about any form of human behavior which enhances the survival of the perpetrator. An economist will not differentiate between the economic activity of a hostile leveraged buyout and the activity surrounding the initial public stock offering of an emerging company. If the numbers are the same, the two activities are considered economically equivalent, even though from an industrial point of view the two activities are as different as death and birth. If the economist does not understand the difference between industrial death and birth, all the numbers are absolutely worthless.

Article of Faith #3: Personal profit at the expense of the common good is virtuous behavior.

INSOFAR AS THE LEISURE CLASS is concerned there is no common good. Resources are limited and the struggle to get a piece belongs to the strong and the shrewd. The winners get to propagate, write history, and be admired. "It's a jungle out there." "It's the survival of the fittest."

It is perfectly understandable that a hunter-leisure class mentality fails to recognize the notion of a common good. The most obvious common good, the biosphere, cannot be damaged except with the help of the industrial class. Predation only leads to the survival of the fittest—an arguable good. Destruction of the biosphere is a leisure class problem only because the industrial class has made the problem possible.

Economists have few conceptual tools to deal with predatory behavior contributing to the problem of a deteriorating biosphere. Whatever problems predatory behavior has caused throughout the bulk of history, destruction of the biosphere has not been one of them. Of course, this is no longer true. Predatory behavior combined with the possibilities offered by industrialization can cause incredible, if not irreparable, damage. If maximizing profits is the goal, how can there be room to trade profits for an improved biosphere? Or any other form of common good? Pathetic attempts to place a price on clean air and water have only demonstrated the economists' dilemmas. At best, the economists can recommend charitable contributions that fit well within the framework of noble generosity.

The industrial class must believe in the notion of the common good by the very nature of their work. Large-scale projects depend on collective action. Not only must there be an intense amount of cooperation among players, but the players must have the minimum number of distractions. If the players are hungry when they work, they will think about their hunger instead of their work. Satisfied, well-fed, and prosperous workers will always do better work than their dissatisfied, hungry, and impoverished counterparts. Technologically simple, though large tasks, like the building of the pyramids or the Great Wall of China, can be accomplished with forced labor. In point of fact, most of the producing class throughout history have been slaves; but, forced labor never produced the artificial heart valve or the microchip. The higher the general level of prosperity, the greater number of potential industrial superplayers, and the greater chance for high technical achievement. Meritorious reward schemes are necessary prerequisites for industrial success, but even the most enlightened industrial enterprises fall strangely silent when the notion of the common good extends to the preservation of the biosphere.

EVEN IF AN ECONOMIST can escape the traps set out by these three very questionable articles of faith, other pitfalls await. Modern corollaries to these postulates mask the hunter roots for a contemporary audience. Consider the very contemporary notion that the success of an enterprise is tied to its profitability: all profitable activities are successes while unprofitable activities are failures.

The problem of assessing the success or failure of an enterprise based on its profitability is that the very concept of what is profitable masks a very great many value assumptions. Not everything worth doing is profitable: not everything that is profitable is worth doing. Raising children, cooking meals for friends, planting trees, and reading books are all examples of worthwhile activities that are almost never profitable. In contrast, selling booze to alcoholics or guns to criminals is almost always profitable but never worth doing.

Moreover, there is the very large problem of defining profitability. The figure a man claims is profit from an enterprise is rarely the same figure if the listener is the tax examiner or a woman he is trying to impress. Stockholders are given glowing reports while union heads hear dismal reports. The more ways an accountant can arrange the same figures, the greater his worth. The only subject other than profits more likely to be the object of creative writing in fiction is profit projections.

Even if some measure of honesty and objectivity could be brought to the subject of profitability, profits are an extremely fuzzy notion because of their definitive predatory nature. Profits are often merely those costs that can be shoved off onto someone else. Many profitable enterprises would become unprofitable if they were forced to pay to clean the water they need and return it to its original state when finished with it. Many operations are profitable because they fail to spend for maintenance: a short-term gain in return for a long term decline. Marx claimed that profits came from paying employees less than they were worth, but inequitable transactions with the planet and with the future are just as important, if not more so, to a working definition of profits. Profitable enterprises that pay all employees living wages, do not degrade but improve the environment, while investing and maintaining the operation for future use, are extremely rare, possibly nonexistent.

The subject of profits inevitably leads to a discussion of money. Economists love to discuss money. Money can be counted and the flow of money can be traced, but discussions of money shroud the greatest of all weaknesses in economics. By the time wealth has been defined in monetary terms, most of the interesting economic arguments have already been decided. Econometric forecasting, the highest flowering of twentieth century leisure class economics, is merely the computer assisted mathematical manipulations of the numbers represented by money. Econometrics is fatally flawed because it tracks money and not wealth. Since mathematics is an abstraction, and money is an abstraction, mathematically tracking the flow of money is a compound abstraction. Using an econometrician's mathematical model to describe the wealth of a nation or any of its subparts is a process akin to reading a description about a set of blueprints for a building and arriving at a judgement about the real building. It is no wonder that the most complex and scientific-looking models of econometricians are about as useful and accurate as a handful of darts and a dart board. Economics passes as a science by virtue of its great powers of quantification. It turns out that counting money does not quantify much.

When the disciple Paul claimed that the love of money was the root of all evil, (1 Timothy 6:10) he was correct. Anyone who can love something as intrinsically nonsensical as money is prone to other forms of psychotic behavior. Money is not wealth but a symbol for wealth. After all, money is nothing more than funny shaped pieces of metal, colorfully engraved pieces of paper, or more commonly on the twentieth century, plus or minus charges of electricity in the core memory unit of some bank's computer. None of these items has much intrinsic value. Gold has some very interesting industrial applications and a handful (truck full?) of paper currency could be burned for warmth, but generally speaking, the value of money has nothing to do with its intrinsic worth.

An extreme but illuminating example of the value of money is shown by the dilemma faced by two men in a desert. One has a jug of water and the other has a five pound bar of gold bullion representing money. The only way the man with the water will exchange for the gold is if he thinks he can have both the gold and his survival as well. He knows a shortcut out of the desert or knows where there is some more water the fellow with the gold does not know about. If, however, the man with the water comes to the conclusion that without the water, he will die, it is very likely that although he would very much like the five pound gold bar, the jug of water means more to him and he will not trade for the gold. Both the man with the gold and the man with the water have come to the same conclusion: in this particular social setting, a jug of water is more valuable than a five pound bar of gold.

This is true even though both men know that just one hundred miles away in another social setting, where water is so plentiful that a gold bar could be exchanged for thousands of jugs of water, neither would even think about trading a gold bar for a jug of water. The social convention is so strong that such a trade would be considered a form of madness. Economists justify the social convention based on a concept of scarcity. In most places, water is plentiful and gold is scarce: therefore water is less valuable than gold, but a jug of water in the middle of a desert is as scarce as a gold bar in settled locations. When money and water are equally scarce, water is valued over money because water has intrinsic value and money does not. In the context of the desert, both men agree that the man with the water is wealthier than the man with the gold. Wealth and money are not the same thing, but so long as social conditions are agreeable, people can be made to believe they are equivalent and convertible.

Those who believe that the person who would take the gold and die in the desert is a fool should consider the current debate over acid rain. In a very real sense, water is being exchanged for money. Some *Wünderkind* in the Office of Management and Budget of the United States, no doubt the same person who came up with the concept of ketchup as a vegetable, figured that stopping acid rain would cost over $6000 per fish saved. Water supports more than the lives of fish and clean water is very rare. Moreover, the nearest source of water, besides the one we have, is a great deal further away than the edge of a desert—the distance must be measured in light-years. Worse than that, the water is not being transmuted into any sort of precious metal such as gold but into a form of money that is nothing more than rearranged electrons in a computer. The environment is an actor in all conversions and failure to take it into account in any social system of wealth convertibility is a serious error.

Money cannot be valuable or worthless. Trading or speculation of money in a commodity market setting has never solved the problem of describing the value of money accurately. In fact, since scarcity or its lack is always an artificial or man-made phenomenon when the commodity is money, the market in money is always more irrational and volatile than that of any other commodity. Make a market irrational and volatile and the market will be corruptible.

Volatile and irrational markets, even without corruption, are essentially anti-industrial. Industrial wealth takes long lead times to create. Arbitrary changes in the rules governing money can and have short-circuited the creation of industrial wealth. If money games cause the real diminution of industrial wealth, the confusion of money and wealth can cause real, but unnecessary, declines in the general standards of living for everyone. Because industrial wealth takes a long time to build and once built, acquires an inertia that can survive some battering, money games exact a serious toll before the money players notice that the real wealth behind the money they have been playing with is no longer there. Industrial distress precedes monetary distress.

Even if an economist abhors the money markets as a system of predatory exploit, the confusion of money and wealth can persist. Wealth, especially industrial wealth, is a subject almost beyond comprehension much less definition, but until the economist understands the nature of industrial wealth, he is stuck with preindustrial models to describe industrial economies. At best, the economist will be able to discuss the social arrangements of how the economic pie is divided, not the much more interesting issues of how the pie got there in the first place, and how big the pie is going to be. From an industrial point of view, this is an exercise in missing the whole point.

NOTES ON CHAPTER FIVE

[1] There is popular confusion in the United States about the Nobel prize for economics. This is understandable because Alfred Nobel did not provide for an economics prize in his will.

NOTES ON CHAPTER FIVE

The economics prize is awarded by Sweden's banking system which uses the prize to advance their social agenda—the rollback of Socialist "excess" in Sweden. Samuelson is probably the most rational of the award winners. The honor was bestowed on him mostly because he sold so many books.

Most economics prize winners can only be described as right-wing nuts—selected to counter some of the left-wing nuttiness of Scandinavian Marxists. In the United States, this Scandinavian cultural conflict is virtually unknown and so the Nobel economics winners often go from utter obscurity in some gloomy closet of academe to respectability—mostly undeserved.

The controversy which surrounds the Nobel prize for economics is equal or greater to that which dogs the prizes for literature or peace. All three trade on the legitimacy of the awards for physics, chemistry, or medicine—which as Nobel had intended, are chosen completely on the basis of merit using objective criteria and without the stain of politics. Americans who assume the same is true for the economics prize are making a serious mistake.

2 In many respects, parts of this chapter are a charitable simplification of a February 1985 *Atlantic Monthly* article by Robert Kuttner entitled *The Poverty of Economics.*

Kuttner contends that economists are operating from premises that are so preposterous, they are forced to argue that the world would be better *if* it really operated as they thought. Their only other alternative is to become intellectual charlatans, a choice a surprising number make. This would be harmless behavior except that many economists believe that they are scientists who are describing phenomena as much a part of the natural order as a thunderstorm. Worse—some people in power actually believe that economics is science and will destroy the lives of millions in order not to disturb these "laws of nature."

Kuttner well describes my feelings upon encountering economics for the first time. Even though I came to roughly the same conclusions in a completely different way, I find his examples so powerful, I could not resist appropriating a few. If you like chapter five, you'll love Kuttner. It is required reading for anyone who has ever taken an economics course.

3 Michael Schudson *Advertising, The Uneasy Persuasion* Basic Books (1984).

At one time, I had planned to devote a major section of this book to the subject of advertising. Advertising is a very interesting subject—any form of persuasion that works is worth examining. John Kenneth Galbraith devotes some space to the effects of advertising on the economy in his seminal work *The New Industrial State.* After researching the subject thoroughly, I came to the conclusion that the real power of advertising was too subtle for a short analysis. Then I read *Advertising, The Uneasy Persuasion.* Anything I could have said about the subject of advertising is better said by Schudson. Of his whole book, I would argue with fewer than 20 sentences. All of his book is interesting, but

NOTES ON CHAPTER FIVE

his chapter eight, entitled *Advertising as Capitalist Realism,* sums up the issues surrounding advertising in the United States better than any other single source.

4 Lester Thurow, *Dangerous Currents,* 1st Vintage Books (1983).

The closest any mainstream economist comes in the United States to an industrial-class economist is Lester Thurow. As part of the faculty of the Massachusetts Institute of Technology, he should be. A regular on Sunday interview shows, Thurow tries to argue issues of industrial competitiveness and his questioners ask irrelevant questions about the stock market. Probably in desperation over this persistent and fundamental misunderstanding, Thurow wrote *Dangerous Currents*, arguably the most devastating critique of the notion of the "free" market ever written.

CHAPTER SIX
MONEY!

"It is well enough that the people of the nation do not understand our banking and monetary system for, if they did, I believe there would be a revolution before tomorrow morning." Henry Ford [1]

MONEY—THERE IS NO SUBJECT that can more deeply divide predators and producers. For the predators, the manipulation of money represents the final bastion of power in a world where they are becoming utterly obsolete. For the producers, money is, at best, a subject of love and hate.

For the first time since the New Deal, banking itself is in serious trouble. The savings and loan business is essentially dead. The Federal Savings and Loan Insurance Corporation (FSLIC) is already gone and the Federal Deposit Insurance Corporation (FDIC) is on the brink of collapse. Major New York banks have seen their credit ratings slide to the point where their bonds are essentially junk.

Many causes are cited—foreign competition, bad loans, inability to diversify, poor or corrupt management, and non-bank lending competition. All of these explanations have some degree of merit, but individually, or even together, these causes do not explain the collapse of a system of lending that threatens to drag the whole economy into depression. The numbers are simply too large.

The cause is much deeper. Banks are failing because of faulty assumptions about money itself. Even though these assumptions about monetary policy are virtually unknown to modern Americans, they were vigorously debated from the dawn of the Republic until about 1940. [2]

The fifty year hiatus in monetary discussions since 1940 prevents most Americans from understanding clearly the problems of banking in the 1990s. Unfortunately, the lack of popular understanding of the issues eliminates the political pressure necessary for a much needed banking reform. This is a tragedy because the current failure of old-fashioned banking provides the perfect opportunity for a critically overdue modernization of the assumptions and practices of lending. Modern banking may be computerized, but the rules and assumptions behind those computer programs are still essentially preindustrial.

An examination of these basic assumptions will show why they are inappropriate for modern societies. Preindustrial international monetary policy has emerged as the key impediment to an industrial-based solution for life-threatening environmental problems.

PREINDUSTRIAL ASSUMPTIONS

1) Economics is about scarcity. Money defines this reality only by being scarce itself. Money has value because it is rare.

2) Charging interest is considered the service fee to the banker for the job of managing money. In the *Kinderspiel* version of lending, a banker is considered a sober citizen who finds money that is not being needed, pays a low rate to induce the owner of this money to deposit it in his bank, lends it to someone else who needs the money at a higher rate, and pockets the difference.

3) Lenders have the right to charge any rate of interest that a borrower will agree to.

4) Small borrowers are risky borrowers and must pay higher rates of interest.

5) Lenders have recovery rights from borrowers. Failure of any enterprise is assumed to be the fault of the borrower (even if the failure is due to natural disasters) who is expected to repay the loan with interest no matter what.

6) Lenders have the right to demand payment at any time even if that action destroys the borrower.

THE NEW REALITY

BECAUSE THE INDUSTRIAL REVOLUTION brought into being methods that vastly increased output, it ended the economics of scarcity for the production of goods. With this new reality, economic arguments would center on the problems of over-production and unemployment. The basic monetary assumption became flawed which, in turn, called into question the validity of all the other preindustrial monetary assumptions.

From the very beginning, the industrialists faced the problem of a money shortage. It should be remembered that Wilkinson, considered one of the fathers of the industrial revolution in England, solved the problem by minting coins himself with his likeness on them. The mere fact that he was forced to create money as well as steel foreshadowed a situation where all breakthroughs in production would spark controversies about the nature of money. Must money be a precious metal? Must it have a noble likeness on it? Could anyone create money? What made Wilkinson's money valuable even though it was quite untraditional? Why should the people who have the power to create money decide which enterprises shall succeed? Why shouldn't the supply of money grow to match the enterprise of a nation?

Gilded Age America saw the introduction of mass-production techniques combined with a deliberate shrinkage of the supply of money. Producers of all sizes were hurt, but it was the farmers of the prairies, who needed the products of industrialization to succeed, who were hurt the most. There were enough of them to form mass political movements around the question of monetary policy.

From an industrial point of view, the most progressive monetary theory of the age was provided by the Greenback Party. The collapse of the party, however, did not end its influence in monetary matters. Charles Macune and Harry Tracy of the National Farmer's Alliance, frustrated by financial interests in their attempts to organize agricultural cooperatives, refined the Greenback theories in their brilliant subtreasury plan. The Alliance found that farmers, like any other producer, eventually encountered the money men—people with the power to ruin any effort. [3]

Monetary theory became in many ways, the dominant political issue of the age. The National Alliance evolved politically into the People's Party (Populists) in 1892. By 1895, the monetary lines had been drawn. The Republican party was wedded to the gold standard, the Populists stayed with their paper currency theories, and the Democrats staked out a compromise position around the free coinage of silver.

From the Populist perspective, the silver position was barely an improvement over the gold standard, but in the election of 1896, a decision was made to fuse their party with the Democrats because of the anti-gold sentiments so eloquently articulated by William Jennings Bryan. The Populist-Democratic fusion went down to defeat at the hands of McKinley and with it, the Populist Party.

Even then the issue did not die. In the election of 1912, the three major parties—Republicans, Democrats, and Progressives—

all had monetary planks in their political platforms. This political consensus led to the formation of the Federal Reserve System by an act of Congress on December 23, 1913.

The new Fed sought to accommodate the need for a flexible and managed money supply as demanded by the old Populists. The idea of a central bank was sold to the congress by Bryan and took shape during the administration of Wilson, a Democrat.

The essential Republican position on the gold standard was kept intact. So was the dominant role of the New York banks. In fact, the Fed was never the compromise it appeared to be though it was an obvious improvement over the chaotic banking practices it replaced. Possibly the strangest feature of the Fed was that while it took over the functions of a central bank, it was never a government institution and remained under private ownership.

Almost immediately, the Fed mismanaged the currency, producing an agricultural depression that began in 1920 and continued throughout the decade. As Bryan had predicted in his "Cross of Gold" speech in 1896, agricultural problems would always migrate to the city. In 1929, the agricultural depression became the Great Crash. As late as 1976, Wright Patman, the longtime chairman of the House Banking Committee, was blaming the Great Depression on the Fed in his comprehensive study of that institution.

The current question that presents itself is: How did such a turbulent political issue die so completely that virtually every American born after 1940 barely understands that monetary policy is even a subject worthy of study and debate? There are really only two answers: 1) During most of Franklin Delano Roosevelt's administrations, the Fed was run by an industrial literate from Utah, named Marriner Eccles, [4] who managed it in the interest of the whole country rather than of the New York banking establishment; and 2) Other government programs of the New Deal such as the Commodity Credit Corporation fulfilled needs left untouched by the monetary reform that produced the Fed. The tradition of Eccles and the modifications of the New Deal produced a post-World War II prosperity that effectively eliminated further monetary discussion except for some residual criticism from the political far right.

The basic flaws of the Fed, however, had been merely papered over. It was still structured around the preindustrial monetary assumptions that have misguided lending since the dawn of money.

Should the leadership of the Fed fall into the hands of anyone who subscribed to the old-time religion, there was absolutely nothing to prevent a repeat of the problems of the 1920s. In 1979, Jimmy Carter appointed Paul Volcker to be the new Fed chairman. Volcker was a preindustrial monetary fundamentalist if there ever was one. The economics of the 1980s immediately began to look suspiciously like the 1920s.

So far, only the structures of the New Deal such as the FDIC have prevented a total 1929-style economic collapse. Unfortunately, as the New Deal structures themselves begin to crumble, the fundamental flaws of the Federal Reserve System are being exposed again.

A new Marriner Eccles could save the United States from a 1930s-style depression. Certainly, that is what everyone hopes for. Alan Greenspan, the current Fed Chairman, is no Marriner Eccles. He is, in fact, more extreme than Volcker. He actually looks to the period between 1873 and 1896 as a monetary model. Those times were catastrophic for almost all the nation's population, and Greenspan's thinking may trigger a catastrophe of a similar or greater magnitude.

The time has come for the rest of us to dust off the monetary response provoked by the thinking he admires so much. Only this time, there can be no doubts as to the nature of industrialization. Maybe it is time to correct the basic flaws in the Federal Reserve System so that loose cannons like Volcker and Greenspan cannot wreck the economy of a whole nation.

Industrial Monetary Policy

MANY ECONOMISTS, POLITICAL THINKERS, AND SOCIAL OBSERVERS have wondered at the contradiction that the great strides in productivity of the industrial revolution have saddled industrial societies with chronic overproduction and unemployment. To a producer, it is not strange at all. Rather it is a matter of preindustrial monetary systems failing to accommodate the potential of industrialization. To realize the full potential of industrialism, producers, like Wilkinson and his successors, must insist on new monetary thinking.

Industrial Monetary Assumptions

Money is only information

MONEY IS NOT A COMMODITY. Money has no intrinsic value. Money is now merely positive and negative charges stored somewhere on a computer chip. The physical manifestation of money has changed over time, but the real nature of providing information has not changed at all. In fact, since computer chips are the heart of the information age, the issue may be easier to understand now than ever before.

Human effort determines the value of money

IN PREINDUSTRIAL SOCIETIES, the creator of things of worth was assumed to be supernatural. People did not create land or jewels. Wealth was gathered or seized.

Farming first proved this assumption to be wrong. There is more to agriculture than the harvest. Agriculture is about planting and tending as well. Those who do the work naturally resent those who assume that growing is only about gathering what God had provided. Even so, farmers supply only a small fraction of the creativity necessary to produce a crop.

As the human creative input became a greater fraction of the finished goods, the assumption that wealth is merely gathered became increasingly false. Take a modern example of a manufactured product such as a microdisc filled with application software. Everything about the product has been processed beyond recognition. The case is plastic which is formed from molecules once found in an oil well. The metal originated in a mine. Even so, a blank microdisc is only worth about two dollars. The rest of the value is contained in the program written on the blank disc. The software writer can legitimately claim that the raw materials necessary to produce his product are less than one percent of the total value of the finished item for sale. If money merely reflects the bounty of nature, it cannot accurately describe the creative value of human input.

Humans have been assigning value to the creations of nature for a very long time. The whole idea of the free market was formed to solve this basic problem. Evaluating the creations of humans is quite another puzzle altogether.

If a new product appears on the market, its value is determined by a formula which looks something like

V=f(HN, CD, T, RI, HE, E),

V is the amount of money validated,

1) **HN** is an assessment of human need whether physical or psychological,

2) **CD** is the creative design input which is a combination of learned information plus the intuitive flash,

3) **T** is the existing technology,

4) **RI** is the resource input from nature,

5) **HE** is the physical effort or work supplied by humans; and,

6) **E** is the energy supplied from nature in the form of fuels.

Because all six elements are critical, many forms of human endeavor do not add to value. For example, the Russian ruble is not worth very much. This is true because the Russians have almost nothing to sell except raw materials. Their manufactured goods are in the main essentially worthless because they fail in so many areas. They are not very creative, they do not meet human needs very well, they do not represent much real work, and are manufactured with primitive technologies. So even though Russian goods are made with prime natural resources and use large amounts of fuel in the manufacturing process, they fail to create value because they fail miserably in the other four areas. Pump more money into the Soviet economy and the result is inflation internally and a devaluing of the ruble internationally. Because of other problems in Soviet society, mere additions of money do not create prosperity. [5]

Suppose a producer goes through all six steps and the money supply does not increase to match the added value. A vicious form of deflation follows. Two outcomes are possible:

1) The new product must displace others in the fight for the available currency. In the beginning stages of industrialization, factory-produced cloth displaced the output of weavers. Most early advances were of this type because the new product was merely a cheaper form of the old product. The result was a massive displacement and unemployment of artisans.

2) If a manufacturer introduces a new product that does not displace an old one and the currency supply does not expand, the product will simply fail. The fight for the share of the currency pie is simply too hazardous. The new producer, in fulfilling the six requirements, has incurred fixed costs. There is a minimum selling price below which a producer simply cannot operate. Typically, the more sophisticated the product, the higher the fixed costs. Sophisticated goods must either sell very widely to justify the high fixed costs, or they must command a very high selling price. Both requirements are extremely difficult in times of monetary stagnation or deflation.

For all its complications, the argument about the nature of money in industrial societies is rather simple. To increase the output of technologically sophisticated goods the money supply must grow. Creative, clever, hard-working, well-educated people must make the new money valuable. Both are absolutely essential. All other questions about money are secondary.

SOME GROUP OR INSTITUTION MUST BE RESPONSIBLE FOR THE ORDERED GROWTH IN THE MONEY SUPPLY

THE ORIGINAL INTENT of the legislation creating the Federal Reserve Board was the simple management of the money supply. Because the United States was then an agricultural country, the great monetary problem to be solved involved expanding the money supply at harvest time so as to cope with the flood of agricultural commodities coming to market. Along with managing the flow of bank checks, seasonally adjusting the money supply is a function the Fed does magnificently.

Because most industries would rather not be seasonal and seasonal adjustments to the money supply are not so important, modern criticisms of the Fed revolve around the old unresolved question of monetary growth. New industrial enterprise requires an expanding pool of new money. Preindustrial banking theories insist that expanding the money supply always leads to inflation. Industrial monetary theory postulates that expanding the money supply generally leads to growth of real economic activity. Both are correct. An expanding money supply leads to inflation in preindustrial activities such as real estate speculation. An expanding money supply also causes industrial expansion.

The Fed is torn between its duty to expand the supply of money and its fear of inflation. The current criticism is that the Fed is a preindustrial relic that worries ONLY about inflation.

John Kenneth Galbraith has argued for years that a different balance can be struck between the two outcomes of an expanding money supply. His observations are taken from the American World War II experience. The money supply was dramatically increased to fight the war. By enforcing price controls, Galbraith's role during that period was to ensure that inflation did not break out in the preindustrial segments of the economy. He was very successful. The American industrial economy exploded into activity during the War, yet inflation was quite moderate. Galbraith's point has been that such a managed economy is possible for any good reason—not only the defeat of Fascism.

When Volcker introduced his preindustrial monetary theories in 1979, producers erupted in rage. Volcker was given Secret Service protection—the first time in history such a move was required.

The producer rage was justified. Why? it was asked, Should producers, who are willing to build a better world, sit idle because the Fed is unwilling to push the few buttons necessary to create more money? A producer creates new wealth, a real estate speculator can only inflate or deflate the existing wealth. Why do the needs of real estate speculators come before the needs of industrial creators? Give the speculators more money and inflation is inevitable. Give producers the money to use and they will produce new riches. Producers hate inflation too—they want inflation under control. Producers insist on another method to fight inflation rather than toying with the money supply. Money is not a game to them, rather it is part of their life-support systems—without it, they die. In point of fact, thousands of producers did indeed die as the direct result of Volcker's foolish experiment.

Someone has to create the new money in an ordered fashion. Someone must also determine which bright-eyed inventor or entrepreneur has a project that will ultimately validate money. In most countries, this job goes to the bankers. The process goes something like this:

A producer approaches a bank for an operating loan. He needs the money to manufacture and market a new product or to increase the production of an existing product.

The banker assesses the creditworthiness of the producer. If the producer is likely to make something with the loan which would validate the new money and has sufficient collateral should the enterprise fail, the banker will consent to grant a loan.

Under modern procedures of banking, banks may legally loan out money at multiples of their basic capital. The total capitalization represented by stockholders and depositors normally ranges between three and twenty percent. The real asset base of a bank is represented by its performing loans.

The banks are allowed and encouraged to create new money. Under United States practices, only the Fed and certain commercial banks are allowed this creative power. It is the practices of these banks that are the focus of producer concerns.

Increasing the amount of money in circulation is often considered banking's primary social function. This is most often done through the checking system. The bank approves the producer's loan, the papers are signed, and new money is entered into the producer's checking account.

The producer spends his borrowed money for any of the six factors.

The producer makes something to be sold at market. Using the money, he repays the loan with interest.

When the banker is repaid, the money continues to circulate only if the banker uses the money to pay himself or his shareholders and the money is spent. The money may also be used to increase the basic capitalization of the bank. In this case, the money that was created when the loan was made is now extinguished—removed from circulation.

What happens to money resting in a bank is the subject of psychological speculation. The producer knows only that money is useless unless it is moving. Henry Ford made money-in-motion the centerpiece of his monetary rage against banking practices.

Both producer and bankers agree this is what happens. Both know that money has been created out of thin air. While producers question the fairness of the whole arrangement, they are certainly convinced that this system is superior to the gold-based monetary systems it replaced.

If money can literally be created and destroyed by keystrokes on a computer, the logical question becomes: when and why does this money become valuable?

The banker's argument is that money gains value through sound fiscal management. By adhering to the capital reserve requirements fixed by the central banks, the banker has prudently maintained the value of money by not creating too much.

A sociologist might argue that money becomes valuable when it is spent and the merchant honors the check. If, however, too much money has been created, eventually the merchant may not honor the check. As a result, the merchants faith in the value of the money in the producer's checking account is an important, but secondary phenomenon—the real work of making money valuable lies elsewhere.

The producer argument is much more convincing. The producer's point is that it was they who made the money valuable. It was they who performed the magic of turning computer keystrokes into a finished product. It was the producer who paid back the loan which allowed the banker to increase the profit of the bank. A producer would point out that even the banker agrees. The purpose of the banker's screening of loan applicants, after all, is to determine which producer is capable of turning money into food, clothing, shelter, or whatever.

Some still argue that wealth comes from the earth because no matter how clever, every producer must have natural resources with which to work. These people have a point, but while all wealth may originate in the earth, what becomes of it is in the hands of the producers. After all, people stepped on diamonds for thousands of years before anyone ever thought of making jewelry and hundreds of years more before anyone thought to make drill bits and phonograph needles of them. Iron ore existed for millennia before anyone made a bridge out of steel.

In fact, the argument "all wealth comes from the earth or is a gift from God" is usually made to discredit the role of the producer and provide cloak of legitimacy to the banker.

BASIC INFLATION

IF THE BANKER picks a producer who does not validate the new money effectively, the new money becomes inflationary—even in a purely industrial setting. Because some producer is always making mistakes somewhere, any society that regularly increases its money supply will have an underlying rate of inflation. Add to this monopolies that can raise prices no matter what else is happening in the economy, and the finite natural order that will not increase no matter what humans are doing, and minor inflation is a fact of modern societies.

INTEREST RATES ULTIMATELY DETERMINE THE EFFECTIVE SUPPLY OF MONEY IN PRODUCER CONTROL

THE PROBLEM OF INTEREST is that more money is being removed from a producer's control than is placed at his disposal. Because of this, money concentrates in the hands of fewer and fewer people. The higher the interest rates, the faster this concentration occurs. New supplies of money only produce a monetary increase with more borrowers or larger loans.

This fact has argued for the abolition of interest over the years, but without interest, there has traditionally been no lending. Increases in the supply of money are in fact possible without lending and an interest-free system of banking is also possible.

Such a radical system has its own problems so these proposals are outside the realm of this book.

INDUSTRIAL USURY

OVER THE YEARS, there has been a huge argument over the subject of usury. In the beginning, usury was defined as any interest payment. For almost 1500 years, Christianity taught that usury was a sin. Since without interest payments, money lending would not happen, Christians turned to Jews for the service until John Calvin came along and made moneylending at interest something Protestant Christians could do. Lending created prosperity, which in turn, took the onus off charging simple interest.

This did not stop the argument over usury—which has now become defined as the practice of charging excessive interest. The question becomes "What is excessive?"

When interest rates are too high, the honest producer cannot validate money fast enough. Only a thief can repay the loan under such conditions. Even if producers do not steal from others, they will be forced to steal from employees, the future—their children, the environment, or a combination of all of them. The problem is that as interest rates increase, the possible profitability of an enterprise must progress from necessities to luxuries to monopolies to criminal absurdities.

From the perspective of a producer, interest rates determine what will be produced:

Low Rates (1-2%). The production of even basic necessities is profitable to the diligent producer.

Medium Rates (3-4%). The producer must switch production to luxury items that have a higher profit margin.

High Rates (5-6%). Only those producers who have achieved a monopoly on the production of some good can pay off the lender and remain in business.

Loan-Shark Rates (7% and higher). A producer stays in business only by engaging in illegal enterprise.

N.U.P.

THE PROBLEMS FACED by an individual producer become more complex when all of society's producers are considered together. If the basic rate of interest is higher than inflation plus growth in the gross national product (G.N.P.), damage will eventually accrue to industrial societies. Since the goal is growth and producers who are successful in validating new money should and must be rewarded, the point where interest becomes usury is slightly less than the growth in G.N.P. plus inflation. This point, hereinafter referred to as the Natural Usury Point, (NUP) would be 6.5 percent if inflation were 4 percent and growth in G.N.P. were 3 percent. Any figure above NUP will cause deflation and other forms of economic distress. Any figure below this rate will cause general and widespread prosperity.[6]

Simple interest brings order to monetary dealings. This is true, if for no other reason than that it keeps the producer on the treadmill. It keeps him honest and provides an incentive to be industrious. The sooner money is paid back, the sooner the returns accrue to the investor. A producer has incentive to pay back today because tomorrow there is more to pay back.

Having said this, it should be noted that interest rates above NUP will provide an even more effective treadmill. The problem is, if the treadmill is run too fast, more than the producer is destroyed.

NUP ONLY APPLIES TO SIMPLE INTEREST

COMPOUNDING INTEREST—the process of adding accrued interest to principal—poses another set of problems altogether. Compound growth of ANYTHING in a finite biosphere is simply impossible.

When someone wishes to describe the glories of compound interest, the story of Caesar's cent is trotted out. The story varies, but if someone had invested a penny at the time of Caesar at five percent compounded interest, by now that investment would have grown to a size where all the pennies in the world, even if the planet were made of copper, would not represent the figure today. The problem is not that one could not convert the pennies into some less bulky form of money.

Banks are not some sort of magic box that grows money at a compound rate. For a bank to have purpose, the growth in money must be a reflection of some other activity in the biosphere. Because the biosphere cannot tolerate geometric growth, no bank can legitimately promise a geometric return on investment. Eventually the geometric promise of the bank encounters the reality that production has not kept pace with expectations. At that moment, the bank becomes insolvent. Compound interest has only one modern purpose—to bankrupt the borrower.

SEVERE 1970S-STYLE INFLATION

THE INFLATIONS OF THE 1970S, the reason given for Volcker's resort to industrial usury, had many causes—monopoly power to raise prices most especially in the area of energy, foolish investments in production of no value (most especially military expenditures), and poorly directed increases in the money supply. The problem was made infinitely more difficult because energy prices, for the first time in American history, were set overseas.

Industrial inflation, because it has many causes as well as manifestations, usually needs a multiple counterforce to address as nasty an outbreak as happened in the 1970s. There are three industrial solutions:

WAGE-PRICE CONTROLS Wage-price controls run up against the precious theories of the free market (another relic of preindustrial thinking brought to us by economists who still believe that hunter-gatherer societies are a valid model for industrial societies.) Free market ideologues hold sway throughout the English-speaking economics profession. These folks would have us believe that because free markets can establish the price of blueberries and fresh fish, they have something to do with the phone bill, the price of oil, or the level of interest rates.

Wage-price controls are only necessary to control inflationary practices of large monopolies. As there are only about 10 labor unions and 500 companies with monopoly power, the technical problems of administering wage-price controls are not very large. The political problem, however, is quite another matter indeed. Those 10 unions and 500 companies have considerable clout. They did not want wage-price controls so they were never seriously tried during the 1970s in spite of Nixon's small experiment.

INDUSTRIAL REDESIGN Oil prices posed an even greater problem. To counteract price hikes set overseas, the only available option to fight the inflation so induced was to reduce energy consumption significantly. Energy consumption is a function of the design of the industrial infrastructure. Energy demand, consequently, is not very elastic. The fuel required by an automobile can be lowered slightly by driving slower, keeping the engine in tune, or maintaining proper tire inflation. Alternatively, people could drive less.

To achieve real energy savings, the fuel-wasting automobile would have to be replaced by one much more energy efficient. To achieve society-wide savings, the whole automobile fleet would have to be replaced. This solution encountered technological, economic, as well as political difficulties. The same problems were encountered in other areas of energy consumption. It is extremely difficult to make a building more energy efficient once it is built. Conventional electrical generation is already at the technological limits of efficiency and represents a huge capital investment. A moral equivalent of war to solve these problems was actually proposed by Jimmy Carter and he was almost hooted out of Washington for his suggestion.

TARGETED TAX INCREASES Raising taxes will remove money from circulation which would counter inflation as readily as removing the money with usury. Raising interest rates is the shotgun approach. Tax increases represent the superior, targeted, rifle approach. Even to fight inflation, there is never a political groundswell for raising taxes. So this solution to the inflation of the 1970s was also rejected.

THE VOLCKER ROLLBACK

WITH ALL THE INDUSTRIAL SOLUTIONS to inflation eliminated, a resort to the old-time religion of preindustrial usury became the only real option left. Sure enough, the criminal level of usury given us by Fedmeister Paul Volcker drove inflation from the system. Actually, the level of usury was not legally criminal because by the early 1980s, a host of states had changed their laws so that Volcker's industrial usury was decriminalized. Some states even had to change their constitutions to make legal what had been illegal since the 1930s.

Why all that legislation was easier than raising taxes, legislating new energy efficiency standards, or creating wage-price controls, speaks volumes about the powerful nature of the banking interests and the old-time religion of monetarism.

THE DAMAGE TO THE ECONOMY WHEN INTEREST RATES EXCEED NUP

THE PROBLEM WITH FIGHTING INFLATION with only monetary deflation is that in industrial societies the cure is worse than the sickness. Inflation is only cured in this way by putting the whole economy on a going-out-of-business-sale basis.

Volcker's version of the old-time religion led to the 1981-82 recession, unemployment, an agricultural depression, and a deindustrialization of American society that meant by 1985, we had lost the international lead in industrial matters and had actually become a debtor nation. Corporate debt, personal debt, government debt, are all logical outcomes when banks raise interest rates above the NUP in an industrial society. It is debt that cannot be repaid. It is the debt caused by a philosophy of usury designed to bankrupt the borrower.

Since 1979, the prime rate of interest in the United States has exceeded NUP by a considerable margin. In fact, the recession of 1991 can be considered a direct outgrowth of excessive interest.

Damage is caused in two ways: when interest rates are far above NUP, or, when interest rates are slightly above NUP but are kept that way for an extended period of time

Stage 1. The first victims of usury are small producers in competitive, credit-sensitive industries such as agriculture and homebuilding. Start-up enterprises must compete in shrinking markets. Most fail.

Stage 2. Existing companies take shortcuts, evade environmental constraints, defer maintenance, cut back on R&D. Wages drop. Layoffs begin.

Stage 3. Social order is disrupted. Financial institutions start taking unnecessary risks because few good producers can pay the returns required. This leaves the fools and charlatans. Layoffs cascade. Governments are stressed trying to cope.

Stage 4. Whole industries begin to fail. Sections of the country are ruined. Homelessness and crime increase. Prosperity, such as is left, becomes further and further removed from the production of goods. Production moves to countries where interest rates do not exceed NUP such as Japan in the 1980s. Trade deficits balloon.

Stage 5. Financial institutions begin to fail. Orderly financial transactions are replaced by speculation and greenmail. Sober bankers become gamblers and crooks.

Stage 6. Governments are bankrupted. Insurance for financial institutions is exhausted. Financial distress becomes nationwide.

Actually, when the above factors are considered, predicting the recession of 1991 was like the prediction of a sunrise—it was never a problem of if but of when.

Never kill the customer

IN INDUSTRIAL SOCIETIES, it is in the interest of lenders that their borrowers prosper. The bankers need the income from enterprise. They have no use at all for an idle factory, a vacant farm, or an empty office building. Bankrupting the borrower only makes sense if a lender wants the collateral such as a bar of gold. The fundamental rule of industrial banking is, "never kill your customer!"

Even when usury is not technically criminal, it is still very stupid. Failure of debtors leads to a failure of creditors. It is not a wonder that as interest rates above NUP prevailed throughout the 1980s, banks got into trouble and the whole debt-house of cards threatened to collapse.

Noneconomic damage caused by interest rates above NUP

THE OBVIOUS MANIFESTATIONS of Volcker's insanity may not be the most serious. Thousands of otherwise valid enterprises failed in the 1980s or were not even started. Kill infant enterprise, then enterprise of the future is snuffed out. The list of foregone enterprises is almost infinite but it includes environmental process and waste management controls, techniques for sustainable agriculture, solar and other renewable energy generation, urban and industrial renewal, and energy-efficient structures.

What we got instead was disaster. Take the sad example of agriculture. Of all the socioeconomic advances in history, none rivals owner-operated farms. Family farms—as they are called in the United States—are so successful because they are self-managing. Farmers who work their own land care for the resource, know what plants grow best, know how each field should be worked, do not have to be told to go to work or be supervised when they work, and best of all, they tend the land as an investment for their children.

When the usury of the 1980s hit American agriculture, hundreds of thousands of farmers, many third and fourth generation on the land, were forced into bankruptcy—only to be replaced by absentee landlords such as insurance companies. Sustainable agricultural practices were abandoned in favor of mining the soil with the methods of mega-agribusiness. Soil erosion and chemical pollution skyrocketed. For what was done to agriculture alone, Volcker should be condemned, but agriculture is only the beginning. The effects of usury can be seen in deforestation and global warming, lawless pollution, and the building of cheap, energy-wasting junk housing throughout the Sunbelt in the 1980s, which ultimately will lead to ozone depletion.

Think of the absurdity of it all for a minute. All the ills listed above, and many others not included, were brought to us by usurers with preindustrial mentalities who accomplished nothing but the reprogramming of some computer chips. Usury did not even benefit the lenders—witness the current crises in banking. Worshipping preindustrial ideas is bad enough, but worshipping misprogrammed computer chips utterly redefines the idea of idolatry.

Recognizing the fundamental problem of an ill-conceived monetary policy as the world careens towards global depression is not enough. Changes must be made soon or the 1990s threaten to make the 1930s look like a picnic. The folks in the 1930s actually had more options open to them than the United States has today. The government was not in debt and so could engage in deficit spending. In those innocent prenuclear days, world war would eventually bail them out. No one today is foolish enough to believe that warfare is good for the economy. Then the United States was a creditor nation and had total control of its economic destiny.

This time, there is no option but to engage in monetary reform. This will take more courage than exhibited by Congress of late. They must rise above their PAC mentality that brought low the Keating Five. The United States government must finally stand up to the usurers.

Item A on the agenda must be a reimposition of usury laws. Never again should the prime rate exceed NUP and money must be created in sufficient quantities to keep the rates below NUP—even if falling interest rates cause a flight of foreign lenders. If this cannot be achieved, then the time has come to nationalize the Federal Reserve System and make the governors stand for election.

Nationalization seems such a foreign concept to Americans, but public decisions of the import of monetary policy are simply too vital to be left in the hands of preindustrial technological illiterates with a plunderer's mentality. Every bank that the taxpayers are forced to bail out in the 1990s should become a public bank—the public has paid for them and they should own them. The argument that public bureaucrats are unsuited to make industrial decisions is utterly specious. Banks have been picking industrial "winners" and "losers" for a long time now. For over a decade, they have been making preposterous decisions. Elected officials can do no worse.

For money to do any good, it must be put into the hands of those with a rational plan for its use—those who understand it as a tool, an opportunity, or a creative medium—not as an idol, a means to power, or a substitute for sexual inadequacy, religious impotence, or personal worthlessness.

Of course, elected officials can have the same mentality exhibited by the monetarist bankers. In fact, most do. The electorate must insist on bankers, public or private, who understand that the industrial revolution happened and further, that they have some understanding of what that means.

Not all interest rates must be limited to the NUP rate—only the important ones—and only on those loans guaranteed by public agencies such as the FDIC. The list of those activities slated for NUP-rated loans is, in fact, quite short. It includes:

Agriculture—especially loans to owner-operated farms and most especially to young farmers who are committed to sustainable agricultural practices;

Renewable energy generation and energy-efficient technologies;

Environmental or "green" technologies;

Public infrastructure such as mass transit, sewage systems, and schools;

Housing—energy-efficient and limited to the primary residence; and

Start-up manufacturing enterprise that makes the pieces for any of the above.

It is certain that if public banks began to write loans at NUP rates for the necessary industrial enterprise mentioned above, prosperity—such as has not been seen in the United States since the 1960s when NUP rates were public policy—would return with a rush. The choice is really between a 1930s style collapse of the American economy and the next golden era.

The ultimate question of the 1990s is, "Are we really going to let a bunch of preindustrial, techno-illiterate, computer-chip idolators march this country down the road to ruin? Really?"

NOTES ON CHAPTER SIX

[1]Robert Lacey, *Ford, The Men and the Machines.* Little, Brown and Co. (1986).

This best-seller is a wonderful read about how Henry Ford produced his industrial revolution. Being British, Lacey finds Ford's monetary notions oddly quaint, but he gets the facts straight.

Ford's monetary ideas are not easily dismissed because he proved their worth. Ford Motor Company grew from internally generated capital—the first such enterprise to grow so large. This was a conscious choice. Ford went bankrupt twice trying to build an automobile company with the "help" of borrowed money and bankers. Had Ford not chosen his unique monetary strategy, there would have been no car for the masses because no banker would have let him try. Fordism IS monetary policy—probably more so than industrial policy. More industries can prosper with Ford's monetary ideas than with his industrial ideas—which as history has shown, have limited applications.

[2] William Greider, *Secrets of the Temple*, Simon and Schuster (1987).

The best and most comprehensive treatment of the Federal Reserve System and monetary policy yet written, Greider is must reading for anyone who wants to understand monetary policy. This book, however, is almost 800 pages long so reading it reflects a real commitment to understanding the subject.

When he describes the historic debates over monetary policy, he is not exaggerating. Greider claims that before 1940, Americans debated monetary policy in barbershops and cafes. Monetary policy discussions were not limited to scholars.

I know from personal experience that he is right. I heard my first monetary discussion from a National Farmer's Organization (NFO) trainer in 1963. Even though I was only 14 years old at the time, the monetary philosophy of the NFO had been so refined for teaching to farmers, that I understood it immediately. For almost 30 years, as a result, I have been an ardent Fed watcher and I have yet to see any of its actions in any way refute the basic NFO monetary position. The world looks very different to those who understand that money is only an idea.

Grieder's position is that the flaws in central banking policy are not the result a wicked conspiracy to control the world, but are honest mistakes of very ignorant people. What makes an American banker ignorant is a lack of awareness of the problems of production. This is certainly not true of all bankers—especially in Germany and Japan. The light went on as I read Greider—Ameri-

NOTES ON CHAPTER SIX

can bankers (and their philosophers, the economists) are not evil, stupid, insensitive, or malicious, they are simply preindustrial. In light of the actions taken by Germany's central bank following reunification in 1989, it was proven once again that if land lust has been triggered, even the industrialized German bankers can suffer a serious relapse of preindustrial thinking.

Living in an industrial country does not guarantee industrial understanding. It may have during the era of industrial formation—though Veblen found plenty of industrial illiterates in his day. In fact, life has been made so easy for the typical urban dweller that it is possible for a majority of the population in industrial countries to never think about industrialism at all.

Unless a person has built something, preindustrial economics makes perfect sense—it was, after all, the only form of economic thinking for thousands of years. Bankers cannot understand the implications of their actions because for them, monetary questions have been closed to discussion for so long, they cannot comprehend the existence of any other alternative.

3 Lawrence Goodwyn, *Democratic Promise: the Populist Moment in America.* Oxford University Press (1976).

This is the definitive history of the Populist movement. The book's sole weakness is that it concentrates on the actual history of the People's Party and somewhat short-changes the immense contributions Populism made long after it had lost its vitality as a third political party. I speak from personal experience. Minnesota and North Dakota were peripheral to the formation of the People's Party, yet I learned populist notions growing up in the 1950s and 1960s—long after the People's Party collapsed in 1896. But even with these tiny criticisms, Goodwyn's book is MUST reading.

Goodwyn's explanation of the impact of a monetary debate on the formation of the National Farmer's Alliance (NFA) is a bit thin but is the best I have found. To understand the debate more completely, research into Greenback Party documents is necessary.

In Minnesota, the NFA's organizer was a former Republican governor and United States Representative turned best-selling utopian writer and orator named Ignatius Donnelly—who would write the party platform of the Populists. As a result, it is safe to say that Donnelly's monetary positions are those of the inventors and builders of the early American industrial forms. Donnelly books are also much easier to find than most NFA sources.

4 John Kenneth Galbraith, *Money*, Houghton Mifflin (1975)

Galbraith thinks highly of Eccles, as well he should. He worked closely with him to ensure that World War II would bring prosperity to America instead of economic chaos—as is so often the case during wars. Anyone who lacks the courage to tackle Grieder's book would be well advised to read Gal-

NOTES ON CHAPTER SIX

braith. The subject of money is arcane—Galbraith makes it interesting and understandable.

5 While it is easy to criticize Soviet society, it should be noted that the United States is now in a very similar position. Were the value of the dollar to be solely determined by the value of manufactured goods, it would have become virtually worthless long ago. As it stands, the dollar has value mostly because oil producing nations price their commodity in dollars. Without that prop, the dollar would be barely worth more than the ruble.

6 To save the reader from doing the math, 1990 inflation in the United States was figured at 6.1% while G.N.P. growth in the final quarter was -2.1%. Subtract the producer 'benefit' of .5% and the NUP for the first quarter of 1991 should be 3.5%.

The magnitude of the damage of industrial usury is obvious—so is the cause. According to the *Fortune* magazine of Dec 1991, the average real interest rate (Treasury bill rate minus inflation) from 1920-81 was .1%. Since 1981, it has averaged over 4%. The wonder is that economy did not crash earlier.

Chapter Seven
Fordism

THE POLITICAL DESIRE of populists to expand the money supply and decrease concentrations of wealth could easily be viewed as a power grab by persons practicing the politics of envy, but populist economics was much more. At the heart was a simple notion. Bad economic times not only idle productive workers through unemployment, but often bankrupts those brave and clever enough to start their own business enterprises. The populists argued that if this terrible waste were eliminated, it would follow that the whole society would become richer.

Arguing the point and proving the point were two very different matters, but out of the economic and political chaos of the late nineteenth century American experiment in capitalism emerged a man whose business philosophies and products would transform the very nature of twentieth century economic reality: Henry Ford.

Ford did not invent the automobile, nor the assembly-line process, nor most of the tools that allowed him to succeed, but Ford's business ideas would prove that he was the Populist's populist. His decision to build a car for the masses would transform the face of America by changing the definition of economic success. Even though Ford merely proved the validity of the Populist's economic agenda, fairness dictates that his economic proof should be honored by naming it for him. [1]

THE FORDIST ECONOMIC PRINCIPLES:

MASS PRODUCTION REQUIRES MASS CONSUMPTION. It is simply not possible to design something as complicated as an automobile if the costs of engineering and tooling cannot be widely spread. Without mass markets, great product sophistication is simply impossible.

MASS PRODUCTION REQUIRES CONTROL OF, RATHER THAN OBEDIENCE TO, MARKETS. The summary judgement of the market—bankruptcy—is inappropriate for large-scale enterprise. Too much is at stake: too much money and too many people. Markets would still have their place under Fordism but significant efforts, including advertising, would be employed to control the behavior of markets.

What made Ford the social revolutionary was his insistence that WORKERS MUST MAKE ENOUGH MONEY TO BUY BACK PRODUCTION. When Ford announced that he would pay $5.00 for an 8-hour work day, he was denounced by his fellow industrialists as a Bolshevik. Ford was no Bolshevik as his later life would demonstrate. He was merely a pragmatist. If people cannot buy back production, the consumer pool dries up—eventually causing failure to the businesses that could not be allowed to fail.

In order for Fordism to work on a society-wide scale, everyone would have to agree with him. Because of the competitive nature of American business, some folks find the need to cheat. In the early days, especially in the heyday of the Model T, Ford was so successful that he could go it alone. He could simultaneously pay his workers the highest wages while selling his product for the lowest price.

Ford had institutionalized a very antimarket notion called value. Price became less a function of markets and more a function of vertical integration, ingenious production techniques, clever design, and resource efficiency. How much should a car cost? This turned out to be a question beyond the scope of any extant economic thought.

By concentrating on value, Ford was able to reduce the price of his product to the point where his production workers could actually buy one. Never before in history had anything so sophisticated been available to almost anyone. It was as if a stone mason were able to buy a cathedral.

Not everyone was in a similar position and by the 1920s, Ford's scheme was under assault. After teetering through the 1920s, the American economy collapsed. John Maynard Keynes provided the economic rationale and Franklin Roosevelt provided the political clout for Fordism's missing link—IN TIMES OF ECONOMIC DISTRESS, CONSUMPTION MUST BE SOCIALIZED.

With this missing piece in place, Fordist economics would dominate the American landscape and would produce what became, following World War II, the Great Prosperity.

How can this be?—an astute reader might ask having studied a great deal of economics and never come across the notions of

Fordism. In spite of its great success, the great irony is that Fordism is far from a commonly accepted form of economic thought in the United States. It is only in Japan and Europe where the term "Fordism" has widespread and specific meaning.

Ford was not an economist and during those times he chose to make statements about how the world worked—outside those areas relating to his immense mechanical genius—he was widely regarded as an ignoramus. A person cannot officially become an economist without writing. Ford could barely scrawl.

Ford invented Fordism by inventing a new industrial yardstick by which every other product would be measured. The automobile still fulfills that function in most industrialized countries. Products either compare favorably or unfavorably with cars in terms of value. Products which show greater sophistication for lower prices than automobiles are considered bargains. Computers and consumer electronics are bargains by Fordist standards while medical equipment, military hardware, and site-built housing are considered very bad buys. Even more importantly, most countries with auto industries use the auto-worker's contract as the scale by which wages are judged.

The American economy became Fordist without anyone seeming to notice. Not surprisingly, this causes widespread confusion. Most professional economists and other economic writers maintain that we have a capitalist economy. In a sense, they are correct because Fordist economics is an economics of production and much of the American economy is non-productive and indeed capitalist. Compounding the confusion, such writers then determine that they must define capitalism using either the eighteenth century notions of Adam Smith or the nineteenth century ideas of Karl Marx. That capitalism should be defined by the twentieth century notions of Henry Ford occurs to but a handful.

American citizens and their politicians are equally confused. The common perception is that capitalism is an epithet hurled at us by our enemies. As a result, the American economic system is commonly called free enterprise or more strangely because it is a political term, Democracy. While calling the combination of Fordism and capitalism these names would have pleased old Henry, it hardly makes for enlightened political or economic thinking.

The time has come for a fuller discussion of the implications of the non-decision to embrace a Fordist economics. While Fordism produced history's most widespread general prosperity, it is hardly a doctrine without problems. Fordism could withstand the concentrations of wealth inherent in industrial capitalism while provided for a comfortable middle class only by victimizing a third party. The third party was the environment. As sentimental as Henry Ford was about the natural environment, his economics and products have almost destroyed the biosphere. For that reason alone, Fordism should be scrapped.

Scrapping Fordism may be a bit premature. Importantly, Fordism is effective. The Japanese have become the richest nation on the planet using Fordist production principles. Even more poetically, The Ford Motor Company, under the leadership of a living, breathing Fordist, became, as of 1986, the most successful automobile company in America—the highest profits, the best value for consumers, and the highest paid workers. By 1991, all American car companies were in trouble, but it can be argued the leadership of Donald Petersen during the 1980s left Ford Motor in the best overall position. Fordism works! Any philosophy with a successful track record should be viewed in terms of modification rather than disposal. There are few enough successful philosophic game-plans as it is.

The question then becomes, Can Fordism be altered sufficiently so that it can coexist with the needs of the biosphere? The simple answer is that there may be no choice. The Fordist effects on the biosphere may be reversible, but the decision to go to a Fordist economics may not be. The problems caused by Fordism may only be cured by a new Fordism.

Moreover, there are political problems that would be caused by abandoning Fordist principles. How does one shrink a middle class without tremendous social upheaval? Even more importantly, Fordist economics is predicated on product sophistication. The environmental problems caused by the production output of advanced technologies are extremely complex. They cannot be solved by people of mere good will. They must be solved by the people who designed the problem. Fordist methods are effective. The Fordist productive middle class has the expertise. The environmental solution lies in redirecting the Fordist capabilities.

NOTE ON CHAPTER SEVEN

1 Michael Harrington, *The Next Left: the history of the future.* Owl Books (1988).

Chapter 2 of his book is called "Fordism." With few reservations, his description of Fordism is highly recommended reading. He is one of the few economic observers who recognizes that Fordism is an economics unto itself. The rules of Fordism in my book were borrowed faithfully from his.

CHAPTER EIGHT

TOOLS!

PRODUCERS ARE AMBIVALENT about money. They love the resources that money can buy, but loathe the time-consuming restrictions. Producer heaven is permanent funding, a modest living, and minimum supervision—such as once a year. Those who made it to producer heaven in the past forty years have given us the microchip, artificial heart valves, . . . examples abound!

Many top industrial companies in the United States pride themselves on how many producer heavens they can create. Companies that nurture industrial creativity include Microsoft, Merck, and 3M (Minnesota Mining). All agree on something important—if you want to sell pet rocks, the development time is short, but if you want something a bit more interesting, it will take a bit longer. Patient money is the Holy Grail of producers because with it, a producer can be the best. High interest rates produce impatient money—it is not that the meter is running, it is that it is running so fast.

Japanese and German banks are notorious for patient capital. This allows the Japanese to extend lifetime employment to top industrial workers in society. The Germans fund extended and very expensive apprenticeship programs, try their best to provide permanent employment and generally organize their society (long vacations, full medical coverage, old-age pensions) so as to minimize the distractions to industrial producers. Both countries define industrial excellence.

The reason producers need patient capital is that new products often require new tools. Tools define the producer pecking order. The best tools make the best products. When patient capital is spent most wisely, it is used to make or buy better tools or to invent better uses for existing tools. Turning money into useful tooling is the defining problem of industrial capitalism.

Producers are *not* ambivalent about tools. They polish and organize them. They scheme to find new uses for them. They dream of owning better ones. They name them. This is not merely true of the motorcyclist who works on his wheels, it is true of the most advanced scientist. The proposed Texas supercollider is but the highest manifestation of tool lust.

Tool lust also serves as a defense mechanism during times of hardest exploitation. Producers have been second-class citizens in most historical epoches, but could always believe that in the end, tools will bring justice. Toolmaking drives the evolution of man.

Man made tools before he was called man. "The hand is the cutting edge of the mind." If the ability to make tools gives dominion of man over beast, toolmakers must ultimately triumph over oppression.

In the Viking pantheon of gods, the sailor-warriors described Odin—the god of courage and cunning, the farmers described Freyr and Freya—the gods of fertility and increase, and the builders shaped the description of Thor.

Thor was patient but hot tempered—not unlike the producer who has worked on a project for six weeks, which then is ruined at the last moment. Thor fought with a hammer. His mind was slow, but deep—much like the producer who has "noodled" an idea for ten long winters before telling another soul about it.

Fortunately, the Vikings never took their gods very seriously. Gods were mostly good-luck charms that folks who traveled about the North Sea in open, handmade, wooden boats could always use.

The secret to Viking society was that all occupational groups had, more or less, equal power. Iceland has already celebrated the 1000th anniversary of her parliament. In the beginning, it was not a perfect democracy and Thorstein Veblen argued that the coming of Christianity in the thirteenth century ruined it for six hundred years. Modern Viking society is very egalitarian—1000 years of practice will tell. The cultural, political, and economic distribution of power between the banking-clergy-nobility bureaucracy, the agricultural sector, and the industrial interests is cordial and fair. Modern Scandinavian societies carefully debate the allocation of power and rewards between these groups.

In important ways, tools spared Sweden the horrors of twentieth century wars—escaping even World War II. Swedish industrial output was so important that a German decision was made not to invade. The Germans feared that what they wanted would be destroyed in the conflict. Ball bearings saved Sweden.

Relying on the power of tools—beating swords into plowshares—is morally challenging in a world that still believes power comes from the end of a gun. Sweden still agonizes over the moral dilemmas of World War II. Making ball bearings for the Germans was not morally neutral nor was allowing occupation troops to cross her land.

Sweden atoned for this by helping Europe to rebuild following the war. From housing for Finland to medical assistance in Greece, the desire to help asserted itself. Even today, Sweden spends more money in absolute terms for emergency assistance than the United

Nations High Commission for Refugees. Tool-driven societies have different foreign-policy objectives than weapons-driven societies. Veblen was correct: great producers bring peace.

Humans can only modify nature in positive and useful ways with tools. The rise of industrialization, including manufacturing and agricultural activities, is a study in the advance of tools. Sophisticated products and mass production are inextricably tied together by the very nature of tools.

THE NATURE OF TOOLS

A PERSISTENT NOTION that has survived into the twentieth century is the notion of hand-made goods. Nothing can be manufactured without the use of tools. Hand-made is only a term to describe goods that are made with primitive tools. Some items, like sweaters and furniture, can be made with primitive tooling and still compete in the marketplace.

Most items can only be manufactured with advanced tooling. There are no primitive options for making a color picture tube and although no one has been heard to brag that his television is hand-made, the television set was assembled with thousands of hand operations. In this context, hand-made and its opposite machine-made, are only rough descriptions of the sophistication of the tooling necessary for production. Understanding the levels of sophistication in tools is to comprehend a very great deal about industrialization. Imagine a continuum with primitive hand tools at one end and the most sophisticated machine tools at the other.

There are two basic reasons why humans have been interested in making sophisticated tools: to *enable* production and to *increase* production. Sophisticated products can only be made with sophisticated tools. A computer cannot be made with a stone axe. The primary producer motivation for increased sophistication in tools is to permit the production of sophisticated products. Peoples who can fabricate sophisticated tools usually dominate peoples who cannot.

The other reason tools become more sophisticated is that simple products can be made much more rapidly as the tools become more complex. Factories displace skilled artisans.

The most interesting fact about tools is that it takes tools to make tools. Making primitive tools with sophisticated tools is a simple proposition. Making a pair of pliers is easy if there is a steel mill and a drop forge. Making sophisticated tools with simple tools is an extremely difficult proposition. The ability to go up the ladder of tool sophistication is the essential story of industrial development.

Hand Tools	**Machine Tools**
Pure examples: Hammer Screwdriver	Pure Examples: Robot welder Punch press
Labor intensive	Capital intensive
High skills required by the primary operator	Low skills required by the primary operator
Extremely versatile Critical for making prototypes	Highly specialized Necessary for mass production
Inexpensive Resource intensive	Expensive Resource efficient
High chance for error Limited product sophistication	Virtually idiotproof High product sophistication
Easily adaptable to change Higher unit costs	Locked into product line Lower unit costs past breakeven
Limited marketing needs	Sophisticated marketing necessary to deal with large production
No two finished products exactly alike	Production runs extremely close to identical
Relatively inaccurate	Extreme accuracy possible (depends on operator skills)
Very fulfilling work	Some jobs are very boring

ARTISANS ARE OFTEN PORTRAYED as victims of this process. Actually, they are the main culprits. It is impossible to go up the ladder of tool sophistication without extremely skilled artisans operating on every rung.

The fact that sophisticated products and mass production are intrinsically linked means that there is a product *business* cycle intrinsically linked to the nature of tools. As products get more sophisticated, the up-front investment of time and resources becomes much greater as it becomes harder to learn how to make a product.

The first stage is product design. This research and development stage is always a money loser and must be supported from external sources. The early market stage is also a loser when production volume is low, mistakes are still being made, and market introduction costs are high.

If the product is a hit, there follows a period of very high profits when only one company makes a desirable product. High profits attract attention, and inevitably, competition. As the competition enters the market, prices fall because the new competitors do not have to spend the money discovering a market for a new product. Falling prices reduce profit margins that can only be made up by increased volume. Every market has its saturation point so eventually increased volume becomes an effort with diminishing returns.

Prices fall dramatically as markets become more saturated and the pressure to reduce production costs become more acute. The shakeout of competition begins. Competition which leads to lower pay scales has the effect of shrinking the potential markets—especially if this phenomena of overproduction is extended to many industries at the same time.

These features would suggest that as product sophistication rises, the product cycle shortens. This is true because tools are in and of themselves industrial products. Almost no product has ever been produced at a profit the first time it was made. The first product is the most expensive and probably contains the most errors. It is inevitable that toolmakers, like anyone else, would wish to profit from their learning experience, which in most cases, represents a considerable investment. But when toolmakers begin to mass-produce production equipment, the possibilities for overproduction become great.

People in business are quite aware of this problem. Producer monopolies are built around specialized tools. The reality is that anything that can be made with mass-produced tools is not likely to produce a profit.

Predator economics tends to focus on sophisticated tools for the effect they have on production output. This is most unfortunate because it focuses attention on quantity rather than quality. Quality improvements do not show up as productivity gains. Quality is extremely hard to count. All things being equal, economists are correct in assuming that people make their buying decisions based on price. Of course, all things are not equal when considerations of quality enter the picture. People make their decisions to buy on a perception of value, not merely on the basis of price. Value is determined by combining considerations of price with considerations of quality. Quality is far too esoteric a notion to enter into the considerations of economic thought. Yet quality as production excellence is the largest determinant of industrial success.

Quality as an industrial commodity is often at loggerheads with prevailing economic thought for an even more substantial reason. By the rules of predation, a successful business practice is to drive down labor costs. This assumes that labor is relatively interchangeable commodity. While lower labor costs would seem to lower the costs of production, this saving is usually a short-term affair that usually marks the end of a product's life. The implications of quality mean that a growing, expanding industrial firm must have the services of talented individuals. Driving down labor costs means driving out high-priced talent, the very talent necessary for industrial success.

The management attitudes of producer types versus those of predatory types are demonstrated forcibly by the attitude toward labor costs. *The producer attitude is to pay employees as much as the enterprise can afford. The predatory attitude is to pay as little as possible.* This may seem a subtle difference and, in fact, may lead to identical pay scales, but in the real world, this subtle distinction is the difference between a growing and a dying enterprise, or between a company with labor peace and one torn by strife. The result is that quality, and more importantly, *the means to quality* have been historically, the primary industrial, producer class concern. [1]

If economists are ever to understand the industrial societies they pretend to describe, they must not only understand the role of sophisticated tools as a means to increase production, but they must understand the role of sophisticated tools as a means to increased quality.

When a predator economist talks about increased productivity, he is referring to an *index of automation*. Industries run by industrial class minds will often make tooling improvements merely for quality improvements—the route to industrial superiority. Fortunately, sophisticated tools can enhance quality and productivity at the same time. The requirements of accuracy necessary for mass production are so stringent that to solve productivity needs, accuracy needs must be addressed. Accuracy is only one-half of quality, however; the other half is design.

THE ECONOMICS OF DESIGN

DESIGN IS USUALLY UNDERSTOOD as an enhancement of visual attractiveness. The process of increasing a product's visual attractiveness is more properly called "styling."

The common misperception of design as styling is understandable—it is what people see. Visual design (styling) has obvious economic impact on clothing or furniture. In these enterprises, visual design is so important that everything else is secondary. Increasingly, visual design has become more economically important in other types of industrial production as well. Automobiles are an obvious example—for many consumers, appearance is the only design consideration when purchasing an automobile.

Appearance, however, is but a tiny fraction of the total design decisions necessary to produce a sophisticated industrial product such as a television. A consumer who chooses a television because it will look good in the bedroom has evaluated less than one percent of all the design decisions that brought the television to the store.

Sony markets a second design feature in its televisions called the Trinitron™ picture tube. This is an example of design as function. Functional design considerations insist that a consumer should be more interested in a Sony because of the way the picture tube is manufactured. The consumer is not directly interested in manufacturing, but is evaluating whether the unique method of manufacture produces claimed picture improvements.

Functional design is very Germanic. Because of the nature of the German language, many products are named with the manufacturing process included in the name itself. For Germans, function *is* design—appearance is a frivolous or secondary consideration. It was a German, Mies Van der Rohe, who said "form follows function." Obviously, if a culture values functional design highly, functional design has economic importance.

Functional design is much more important than visual design from a production standpoint. Depending on the product, function is 20 to 50 times greater a design problem than appearance. Consumers buying cars are usually unconcerned about the ease of muffler replacement, and so on, but if the manufacturers are clever, this is very important to them. If muffler replacement is very easy, the dealer will be more profitable, inventories can be less, and mechanics need less training. On the other hand, if the job is difficult, specialized tools must be invented and so on.

Even functional design is not ultimately the most important. Mies said "the Gods are in the details." Design problems increase exponentially as a function of the number of parts and manufacturing steps. From an environmental and economic standpoint, the most interesting design decisions are production designs. It is here that economics is just beginning to understand the role of design and where conventional econometrics utterly fails.

Nucor Corporation makes steel in Crawfordsville Indiana with .6 man-hours per ton. The industry standard is 14 hours. Do Nucor workers work 23 times as hard as Bethlehem Steel workers? Of course not! [2]

In 1986, Ford introduced a model called Taurus. Built in the Atlanta factory, Ford required 54 man-hours in 1986 to make a Taurus. By 1991, the time had been reduced to 17 hours. As the assembly-line rates were increased by only 10 percent during this time, other factors were much more important.

Production design is the difference, in most cases, between a company or product succeeding in the marketplace, or failing completely. Thousands of tiny decisions from the layout of the factory, the tools, the assembly scheme, materials selection, inventory methods, delivery schedules, and so on, ultimately determine whether a product can be made inexpensively enough to meet a price target. For the Marxists who believe in labor value, it must be remembered that little if anything about production design is about labor exploitation. This is not Taylorism!

The Nucor steel example demonstrates the concept of a technological breakthrough. Since 1847, steelmakers have tried to perfect a method of continuous casting. After over a hundred years of determined tinkering, a German firm perfected a mini-mill that achieved steelmaking's Holy Grail. Nucor was the first customer.

The Ford example demonstrates that hundreds of correct little decisions can equal a technological breakthrough. Ford did not develop the techniques—an engineer for Toyota named Kenechi Ohme organized detailed production design concepts such as Kanban, *poke-yoke*, and robotic pre-design. *Poke-yoke* means redesigning tools so that everything is "idiot-proof." If a part can be installed upside down, change the design so it can only be put on correctly. [3]

It was found in most cases, that by designing an assembly process so that a stupid robot could do the job, the job ultimately could be eliminated and the robot not purchased. A car bumper is a car bumper. The consumer does not care if it is assembled from 200 parts or 10; the consumer only cares about cost. If the 200-part bumper takes 2 more hours of assembly time than the 10-part bumper, it must cost more.

Along with incredible productivity increases (as measured by simple output per man-hour), lean production makes much higher quality possible. Toyota's Mercedes-beater, Lexus 400, had initial quality reports of less than one defect per vehicle. This is a VERY complicated car with thousands of parts (because an average car contains 13,000 parts, the big Lexus is estimated to have more than 20,000). Yet final assembly takes less than 18 man-hours without needing inspection. Defect rates of 1 per million are possible in some operations. A Lexus proves the economics of building a product correctly the first time. This form of production excellence is not a function of careful craftsmanship, it is a function of design.

DESIGN AS A GROWTH FACTOR

IF QUALITY IS A CONFUSING notion that economists would rather not deal with in their calculations, imagine their confusion with the concept of design. In fairness to the economics profession, some M.I.T. economists have actually attempted to quantify the economics of hyper-lean production in a book called *The Machine that Changed the World.* They are to be congratulated for even trying! If quality is a very difficult notion to quantify, design is impossible. Yet design is at the very heart of industrial success or failure and any description of an industrial economy must come to grips with the importance of design. Economists can describe the process of how the economic pie is divided without understanding design but cannot describe how the economic pie got there in the first place.

Because design decisions are so important, design know-how is an industrial asset. Which endeavors produce a commercially important product and which produce garbage is a function of design. Since it takes, in most cases, the same or *less* labor, energy, tooling, and natural resources to execute a good design as a bad one, it logically follows that design is the determinant of industrial success or failure. Design is the difference between *wasted* and *productive* effort. Wasted effort implies wasted resources and since the natural resource pie is finite, wasted effort means the available pie has contracted. Correct design decisions are the difference between and expanding and contracting economic pie. Industrial economic growth is a function of design.

Of course, design in industrial economics is not limited to simple things. When real energy prices stopped falling and began to rise in 1973, it became very clear that America had made some preposterous *design* decisions. From a design standpoint, everything was obsolete in the sense that everything was environmentally unsupportable: the cities were too spread out, the buildings used too much energy, the transportation system was addicted to liquid fossil fuels, and agriculture was based on petrochemicals for power, fertilizer, pesticides and herbicides. Millions and millions of design decisions that appeared to be correct when they were made, were now glaringly and demonstratively wrong.

Since design is placed outside economic thought, design solutions were almost never considered in trying to solve the economic problems that increased petroleum prices brought. Economists are loathe to talk about computer-designed, vegetation-generated microclimates around dwellings, for example, and the effect they would have on discretionary spending. Design has an economic impact but it is not considered to be economics.

By leaving design considerations out of economics, predator economists have demonstrated their *preindustrial* mentalities. Economics without design considerations may describe something but clearly it is not describing industrial societies.

PRODUCER VERSUS PREDATOR CAPITAL

PREDATOR ECONOMISTS are often quite careful to point out that money is not capital in their writings, but no economist actually seems to believe that money and capital are quite different and only occasionally convertible. An economist talking about capital formation is talking about raising money using an array of financial instruments; not about patents, tools, skilled workers, factories, educational institutions, and the rest of the industrial infrastructure that is really capital. Real capital is the subject of textbooks. Capital as money is the subject of the world on Wall Street and in the halls of Congress. Economists give lip service to real capital but fashion their models around *money*.

Every product that is manufactured for sale must answer some significant questions before it becomes an industrial success. The number of questions is infinite because there is always the nagging doubt that the vital question was unasked. These questions are of four basic types.

QUESTION #1: WHAT IS TO BE PRODUCED?

What is to be produced is usually solved by industrial creativity, often called inventiveness. Inventions, as perceived by patent offices, are not equal in importance, marketability, or effectiveness. Penicillin is more important than twist ties. A large percentage of inventive activity is not found at the patent office because it remains a trade secret. Industrial secrecy retards industrial growth because technology is cumulative. Inventions and trade secrets are not the only elements in the decision of what to produce for sale.

QUESTION #2: HOW WILL THE PRODUCT BE PRODUCED?

Economists are in love with the concept of productivity. Simply put, a worker is more productive when making 200 widgets a day than if making but 100. Early industrialization may be described in that manner. The textile mills were good examples of how technology made it possible for each worker to put out more goods. Sophisticated tooling was not the only method for improving productivity. Once the tooling was in place, work could be speeded up. The workers could be made to work faster and longer. Most economic arguments are about how to raise productivity and how to divide the proceeds.

Strangely enough, though political and economic arguments have raged since the dawn of the industrial revolution over the issues of productivity, most arguments are beside the point. Productivity, as measured by output per man-hour, assumes the settlement of some very interesting industrial questions. PRODUCTIVITY is merely one measurable form of INDUSTRIAL EFFICIENCY. If the goal is to make and sell the most desirable good for the lowest possible cost, other far more economically interesting forms of industrial efficiency must be brought to bear on the problem.

There is DESIGN EFFICIENCY. The quicker an enterprise gets to a cost-effective solution, the less has been spent on development and intangible costs. Designing a cost-effective solution involves other forms of efficiency. Effective use of resources is a primary design goal.

RESOURCE EFFICIENCY at an oil refinery means that a method that gets 30 gallons of gasoline from a barrel of crude using 100,000 kilocalories of energy for the process is more efficient than one that gets 29 gallons of gasoline using 110,000 kilocalories.

CAPITAL EFFICIENCY is important because it measures how effectively money is converted into the means of production. Equal amounts of money can have unequal outcomes in the conversion process.

QUESTION #3: ARE THERE ANY CUSTOMERS?

Anything made for sale has assumed a customer from the start. The assumption of customers is based on assumptions that range from a hunch to extensive market research. (Critics of market research maintain there is no difference.) Whether there are any customers when a product gets to market is the big question. Fickle customers and industrial competition for those customers are acceptable hazards of manufacture. The unacceptable hazard, the one that angers the producers more than any, is the final hurdle over which the industrial class has seldom had control.

QUESTION #4: DOES THE CUSTOMER HAVE ANY MONEY?

No matter what is made, how well it is made, or how cheaply it is made, a product cannot succeed if there are no customers with money to spend for the product. Customers have the least amount of money to spend during economic downturns. Periods of economic downturn, whether they are called panics, recessions, or depressions, have as part of their nature the destruction of productive capacity. Predator economists go as far as to say that the silver lining in the cloud of economic downturns is the elimination of excess productive capacity. The same thing, incidentally, is said about warfare. The elimination of the weaker producers allows for more prosperity for the remaining strong producers. The strong producers, it is argued, are making a superior product so periodic culling of the industrial society leads to stronger, more efficient producers and eliminates those producers who should be doing something else.

There are two flaws in this form of reasoning: One is what to do with the people who are thrown out of business by economic downturns. There are humanitarian and social considerations that must be addressed. Are the displaced going to be supported by the rest of society, and if so, at what level? If the level is high enough, would it not be better if these people were doing something productive rather than nothing at all? If the level of support is very low, are not the successful producers running the risk of terrible social upheaval as the gap between subsistence and affluence widens? How rich can the efficient be at the expense of the rest? Is it really possible to be rich while one's neighbor is poor?

The second and greater issue is that all weak producers are not bad producers. Excellent products can fail if their introduction coincides with the onset of an economic downturn. Ongoing successful companies can collapse when a new product fails with the onset of recession and brings the rest of the firm down with it. A serious economic downturn is simply too crude an instrument for culling out bad producers because it takes good producers along with the bad.

Although traditional economic thought would lead us to believe that lower prices are always good because they enlarge the market by making goods available to more people, this assumption is intrinsically flawed. In order to remain in business, industrial producers must have a price high enough to cover fixed costs. It is in the *buyer's* interest, surprisingly enough, that prices do not fall too far. Buying any sophisticated industrial product, such as a car, stereo, or appliance, from a liquidator is risky if the producer has gone out of business. Buying a $10,000 car for $5,000 may seem a terrific bargain until it is realized that the failure of a special $10 part can render the whole car useless. Part of the price of any item must be the continued health of the producing industrial enterprise.

Veblen pointed out the main predatory instrument of the leisure class is the irrational business cycle. Industrial growth is a long-term thing. If an invention takes 13 man-years and the work is being done by one person, the work has to be financed for 13 years. That is long enough to cover 2 or 3 business cycles. Veblen's point is that it is in the interest of the predators to have booms and busts, while it is imperative for the producers to have conditions as constant for as long as possible. It was for this reason that Veblen called Wall Street speculators "industrial saboteurs." Veblen was right, of course. The greatest period of industrial expansion occurred immediately following World War II when international exchange rates were fixed, interest rates moved in very narrow ranges, and labor peace was bought in the industrial nations by giving workers a wage high enough to have some discretionary income. Widely floating interest and exchange rates, severe swings in the business cycle, and the lowering of the basic living standards should, according to Veblen, result in industrial confusion. Industrial confusion pretty well sums up the decades of the 1970s and 1980s.

THE JUSTIFICATION FOR GREEN DESIGN

IT IS PRODUCTION DESIGN DECISIONS that are most important to an industrial-environmental solution. If a factory is thought of as one big tool, then tool design must cover every step from the moment raw materials are extracted from the biosphere until they are returned. If a new green mandate is introduced into the production design process, it may be possible to live more gently on the earth.

The Ohme-Toyota production concepts are now universally known to anyone remotely interested. Dozens of books have been written on the subject. Because it is possible to make everything perfectly, the design question moves toward environmental design by moving back toward functional considerations. Production perfection means that products can differentiate themselves in the market only by function and features. If everything is manufactured equally well, what will motivate the consumer to buy one product over another?

Producers hate such a situation—their nightmare is for their product to become a commodity. Then they must appeal to consumers only by price. Product differentiation by appearance is a problem with mature products. The alternative is to make environmental considerations a product feature. This is what makes green design attractive from an economic standpoint.

Green design is not a pipe dream. All 1992 cars made in Germany must be recyclible by law—even the Model-S class Mercedes-Benz. Volkswagen has established 20 recycling centers. The 1991 Frankfurt and Tokyo auto shows stressed environmental issues over quality or performance issues for the first time in history.

In some markets like Canada and Germany, this fact is displayed prominently in all their marketing literature.

Now it is true that $120,000 gas-hogs claiming to be environmentally pure may be a bit much for most environmentalists to swallow—there is more to environmental purity than waste reduction. Yet, it is a beginning that proves the possible. Mercedes performed a useful service. By lending caché to upscale green marketing, the company is showing everyone else how it can be done.

When the claim is made that pollution is a function of design, the design consideration in question is about 75 percent production design and 25 percent functional design. Elegant technology is simply successful product design combined with environmental sensitivity. By this definition, elegant technology is already economically possible and partially successful. Products can be built perfectly to any design. The immediate goal must be to ensure that the next wave of new products is designed with the good earth in mind.

NOTES ON CHAPTER EIGHT

1 Quality as a function of tooling cannot be understood if quality is measured by the selling price of a product. For example, it would be difficult to conceive two products more dissimilar than a Toyota and a Rolls Royce even though both are automobiles. The Rolls is very expensive, costing 10 times what a Toyota costs—or more. The Rolls is made in limited numbers while the Toyota is the soul of mass production. Rolls Royce has made fewer cars in its over 75 year history than Toyota makes in a month.

Because people often associate quality with a high price, the natural conclusion would be to assume that the Rolls is a technologically superior product. Such a conclusion would be wrong. The engine in the Rolls is of a design over 30 years old. Rolls Royce has not had the resources to design a new engine even though it is outdated and needs replacement. By contrast, Toyota has introduced 10 new engines in 10 years. Rolls must buy major components such as automatic transmissions and air conditioning systems from outside suppliers (in this case, General Motors.) Toyota designs everything in house. The Rolls is not particularly enjoyable to drive because driving is the work of a chauffeur. It is slow, ponderous, and inefficient. Operating costs, ease of maintenance, and fuel efficiency are considered irrelevant. The Toyota is faster, more reliable, cheaper to maintain, more rust resistant, and a great deal more fuel efficient. It even has better seats.

The difference is not one of intent. Both automakers strive hard to make an excellent vehicle. The difference is the resources associated with mass production and sophisticated tools. Toyota has an army of engineers able to attack the smallest problem. Rolls Royce relies on skilled labor to work around production problems. Toyota believes that with enough design effort, every assembly operation can be made idiot-proof. The Rolls is a magnificent automobile even if only a monument to primitive technologies.

Ferrari, another expensive car of limited production, was even less sophisticated than Rolls in its early days. The left and right sides of early Ferraris are not mirror images which gives them a personality not unlike a human face. Only the purchase of Ferrari by Fiat saved Ferrari from becoming the perpetual "kit-car" whose attractions were a jewel-like engine, a barbarous top speed, limited production, and high cost. These were supposed to be enough to make an owner forget that the door handles fell off. Does this mean that an inexpensive Toyota is a better car than an expensive Rolls or Ferrari? By any objective standard, the answer is yes. The reason is that when it comes to product sophistication, large companies with huge design staffs can simply overwhelm small companies with fewer resources. These business leviathans have it within their power to set the industrial standards. The Japanese have set the standard for a sophisticated, cost-effective automobile. Such size and strength comes only from mass production and marketing. Toyota has also proven that once it is learned how to make a cheap car better than an expensive car, making an expensive car better than anything on the planet is only a matter of time.

Those who dislike large companies must face the fact that without large companies, a large segment of the industrial reality would simply not have happened. The value of the telephone is not in the cost of the handset, it is in the number of people that can be called. Television would not exist without mass production. A $250 television would cost $250,000 if a company only made 100 per year (even if such an unlikely company were possible.)

[2] Richard Preston, *American Steel: hot metal men and the resurrection of the Rust Belt.* Prentice Hall Press (1991).

This book describes the characteristics of the men who first built and operated the continuous casting steel mill. While this will hardly resurrect the Rust Belt, it is a stunning achievement. It is tempting to cheer the can-do spirit of Nucor until it is realized that the casting machinery must be imported from Germany where it was invented and perfected. The American cowboys are confronted by the reality that the Germans treat this sale as merely another installation in an underdeveloped area of the world. Anyone who doubts the magnitude of American deindustrialization will find this book sobering.

[3] The ultimate way to understand the role of design in industrialization is to consider the recipe. It is not only food preparation that involves a recipe—everything made by humans has one.

For example, near every wheel on virtually every motor vehicle in the world is a simple item that Americans call a **shock absorber** and the British more properly call a **damper**. The job of a damper is to damp out the natural

NOTES ON CHAPTER EIGHT

oscillations of a spring. Once a spring is set in motion, it will remain in motion until the internal friction of the spring causes the motion to cease. Without dampers, a car that hits a bump will bounce up and down like a pogo stick all the way down the block. Not only is this very distracting to the driver and can cause motion sickness, it is very dangerous because the wheels can actually bounce off the road. Terrific tires, superb steering mechanisms, and brilliant brakes are all absolutely worthless any time the wheel is off the ground. Dampers allow the springs to absorb the shock of a bump and return to their original position with a minimum rebound. No bouncing: no wheels flying off the road: no driver getting motion sickness. No question about it, dampers (or shock absorbers) are very important parts of an automobile.

Like most things important, dampers are extremely unglamorous. They operate close to the road in filthy, hot environments. They take a terrific beating. The life of a damper is not easy. Dampers wear out and must be replaced. The average life of a damper is about 3-4 years; shorter if the vehicle travels unpaved roads or New York streets; longer if the vehicle is lucky enough to reside in an area where the roads are maintained. There are 150 million vehicles in the United States with at least 4 wheels and theoretically over 100 million dampers should be made just for replacement purposes every year; not to mention the 50 million or so for new vehicle manufacture. Dampers may be unglamorous but economically they are a very big deal; 150 million of anything is a large order.

Even though dampers are simple in concept, good dampers are very sophisticated. Basically dampers work by sliding a piston up and down in a cylinder filled with oil. Oil is forced through an orifice from one side of the piston to the other. How much damping effect is produced by the damper is largely determined by the size of the piston and the size of the orifice. Changing the size of the piston changes the production cost of the damper. *Changing the size of the orifice does not change the cost of the damper. For all practical purposes it costs the same to drill a .345" hole as a .362" hole.* The person who decides the size of the orifice has a pure design problem: there are no accounting complications. Pure design problems are those problems which can cause the difference between product success and failure, yet the only evidence of a solution is the correct design specification. A pure design solution should be thought of as an industrial recipe. When the subject is the size of the orifice, the only consideration is the correct answer.

Dampers are part of a larger subsystem of a vehicle called the suspension. Designing a suspension is a very complicated process using calculations involving the suspension geometry (the path the wheel takes as it bumps up and down), the weight of the tire, brakes, etc. which must be controlled, the weight of the vehicle, and the spring constants. Every one of these factors bears directly on damper design. Designing a new suspension takes thousands of hours of computer time weighing these factors against each other followed by equal efforts in testing preproduction prototypes. This is not as scientific as it sounds because the design goals are as subject to fashion trends as the rest of the vehicle. Not all goals can be satisfied. A damper designed to provide a silky ride

NOTES ON CHAPTER EIGHT

over rough roads cannot provide the control for high-speed driving. The dampers on a Buick and an Indianapolis 500 racer are not interchangeable. As roads get rougher in industrial countries, suspension philosophy changes. The trend in automobiles is towards softly sprung, heavily damped suspensions because roads are getting worse. The French automaker, Peugeot, makes arguably the best dampers. They claim their dampers are so good because their extensive experience in Africa with terrible roads has given them a head start in designing a car that can negotiate the Dan Ryan Expressway in Chicago.

Not only are damper designers important because their industry is so large, they are important because they are part of the *fashion* segment of automotive design. If the damper designer specifies the wrong orifice, the car can become an ill-mannered, dangerous vehicle. But if the damper designer solves the problem of the orifice size (as well as other problems) very well, the car becomes such a sweet-handling device, that people will pay *extra*. Good damper design not only sells dampers, it helps sell the whole car. Yes indeed, in an industrial sense, simple, *production cost-free* considerations can have a very large economic impact, but while there may be no production cost considerations, either labor or resource, there are serious design cost considerations. Design efficiency can be measured by how quickly the ideal solution is reached. A design solution which works on the first attempt is obviously cheaper than a solution which has taken 1000 attempts to reach.

CHAPTER NINE

THE GREAT RIVERBOAT GAMBLE

HOW TO LOSE IT ALL IN ONE THROW
THE CURRENT INDUSTRIAL DILEMMA

THE WORLD DID NOT get into its current economic problems by lack of attention. Serious efforts were made to head off problems. The economy has been the top topic of discussion in nearly every country in the world. Yet, every solution tried seems only to make the problems worse. The economic problems of the 1990s are an extension of the absurdities of the 1980s.

Just when economists in the industrialized democracies thought that the problems of managing an industrial economy had been reduced to a matter of fine-tuning in the 1960s, a serious episode of inflation broke out. As we have seen, the inflation of the late 1960s-early 1970s had a typical cause: the Vietnam War. The negative economic consequences of warfare have never been learned in the United States Until Vietnam, warfare had been very good to the United States: there were no negative lessons to learn.

Before the the mid to late nineteenth century, warfare was far less destructive. Professional armies in the service of some noblemen got together at some agreed upon field of battle and had it out for a few hours. Warfare was a romantic, if unpleasant, way to die. Twentieth century warfare changed all that. Mass production produced mass warfare; soldiers became less important as targets. Military minds reasoned correctly that the industrial base of an army was as important as the army itself. Factories and the cities around them gradually became targets. The road to total warfare led to Hiroshima. Compared to the costs of destruction, the costs of production for warfare were rather unimportant. The equation changed in Vietnam.

Vietnam was unique in history in that it pitted an army representing the glories of industrialization against a revolutionary army with almost no industrial base. True, the revolutionaries of Vietnam were supplied from industrialized nations such as China and Russia. Even so, the gap in weapons was simply astounding. More interestingly, the lack of an industrial base for the revolutionary army confused United States military minds who had been trained to fight against industrially based armies. Lacking conventional industrial targets, the military was faced with the problem of redefining their objectives after entering the battle. Failure to define a victory or any other credible military objective doomed the United States involvement in Vietnam.

The United States spent $150 billion on the war in Vietnam and came away with nothing to show for it. It was not the money but the resource realities that the money represented that would eventually lead to the economic problems of the United States The Vietnam war was an industrial war. The $150 billion represented a great deal of petroleum consumed, a great many resources wasted, production capacity diverted from civilian competition, and productive minds wasted on destruction instead of addressing the growing industrial crises.[1]

This was not how the United States remembered warfare. World War I had marked the change in status of the United States from a developing debtor nation to an industrial creditor nation. World War II was an even bigger bargain. The American dead from World War II represented less than 1.5 percent of the total. The United States emerged from World War II with an intact and highly productive industrial plant while Europe lay in ruins from Moscow to Rotterdam. England gave up on trying to be an imperial power and more or less willed what remained of her empire to the United States Best of all, because of the Bretton Woods agreement on monetary policy, all international postwar reconstruction activity would be commercial transactions in dollars. To get so much for so little skewed the thinking of a whole generation of political leaders on the subject of warfare.

The prosperity of the post-war period contrasted very well with the misery of the years preceding World War II. It was not the New Deal that pulled America out of the depression, as economists have argued, it was the economic stimulus of the War. Warfare brings great gains at bargain prices and warfare cures depressions were the lessons learned by economists in America. Whatever the costs in human lives, warfare is good for the economy.

Vietnam should have taught the economists of the United States that warfare was only a good investment if you won. Vietnam taught them nothing. The manufacture of arms represents real, measurable, economic activity. The arms business creates jobs and tools and factories. This activity, however, is a negative economic factor unless a war is fought and real gains are made. Occasionally the manufacture of arms produces a new industrial spin-off that may have commercial value: but these occasions are very rare and never justify the investment. Anything that can be developed for wartime use can be developed under civilian auspices for less. Unless something valuable is seized with the help of the weapons made, all wars and their preparations are pure waste. Waste is by definition a negative economic influence. Waste is the way the rich become poor.

President Lyndon Johnson wanted to fight the painless war. Lyndon Johnson also had a social agenda to wipe out poverty in America. This would cost money. He refused to raise taxes for the Vietnamese war believing that a request for increased taxes would sink his social agenda. Vietnam would be fought on credit. People rising out of poverty have money to spend. Borrowed money for warfare puts additional money into circulation. Add them together and demands for goods and services increase. Real production is diverted to the uses of war. Demand exceeds production and you have inflation.

Wars generate a strange mentality. When young men are dying for whatever reason their country is offering, it seems tacky and tasteless to suggest that they are wasting their time and everyone else's money. Simply dying is not enough to cause such sudden concern for lives lost: more people have died at young ages building America than have died defending it but dead coal miners, construction workers, truckers, or farmers have never been cause for much concern. To suggest that the Vietnam war was less than a noble effort became fashionable as the war ground on and the insane levels of brutality became known. But to suggest that the war was causing economic problems flew in the face of both patriotism and the experience of World Wars I & II. Any economist who suggested as much was branded as a traitor and a heretic. As a result, any economic suggestions of the Vietnam era were usually confined to criticisms of Johnson's social agenda.

As the inflation of the Vietnam war gained momentum, it was aided with the run-up of energy prices that continued the inflation after the American involvement in Vietnam ended. Real industrial inflation was addressed with leisure-class thinking. A brainchild of Alan Greenspan—the future chairman of the Federal Reserve Board, WIN buttons (which meant Whip Inflation Now) appeared on the lapels of famous politicians including the President. Public relations campaigns against the inflationary psychology were run on television.

Nothing seemed to work. A decade of inflation severely altered the economic landscape. Increasingly radical solutions for the inflation problem were suggested during this time. By 1979, one of the most radical solutions was being taken seriously. Its most prominent voice, Milton Friedman, had been awarded the Nobel Memorial Prize for Economics—a certain stamp of "quality." The solution was called monetarism. Jimmy Carter appointed a monetarist named Paul Volcker to head the Federal Reserve Board and the experiment in monetarism was under way.

It is impossible to overstate the importance of the changing assumptions about money. To move a steel factory from one spot on the globe to another requires months of planning, complex logistical coordination, huge transportation facilities, and massive transfers of human potential. To move $500 million from London to Hong Kong requires but a few keystrokes on a computer. One operation requires months or years to execute, the other seconds. There is no way for a producer to escape international monetary considerations—he can never change operations faster than the movement of money.

Because of the role played by the dollar in post-World War II reconstruction, dollar-denominated bonds have been issued in most countries of the world. The result is that a monetary assumption in New York rapidly becomes official doctrine in Argentina or Belgium—monetarism in the United States became monetarism internationally. Those countries that for political or social reasons chose not to embrace the monetarist fashion quickly enough, found out that the new religion was not a matter of choice. When the time came to refinance their debts, the International Monetary Fund enforced the new assumptions on all doubters. Only Japan's central bank refused to bow to the pressure because only Japan had the power to resist. Everyone else eventually caved in. Governments were reduced to the role of managing decline and determining who would get what slice of a shrinking economic pie.

Monetarists believe that inflation is caused by too much money chasing too few goods, but monetarists have no interest in the production side of the equation. They believe the solution to inflation rests in holding down the supply of money, not by doing anything that would increase the amount of goods. Volcker shifted his focus from setting the supply of money so as to keep interest rates in line to a strategy of holding the growth of money to within preset targets. Keep the growth of the money supply low enough and eventually inflation will be driven from the system. When this strategy was implemented, interest rates soared. High interest rates, combined with the humiliation of the American hostage situation in Iran led to the defeat of President Jimmy Carter and elect Ronald Reagan. Reagan arrived in Washington with an economic agenda even more radical than monetarism. Reaganomics, as it came to be called, combined monetarism with something even more preposterous: supply-side economics.

When Reagan arrived in Washington, the country was in the mood to try anything; even if it was a flawed idea. Monetarism was already in place so Reagan only had to put his version of supply-side economics into action. Supply-side economists proclaimed that the industrial weaknesses in the American economy were the fault of insufficient investment. Investment had been diverted to speculative idiocies such as diamonds, collections, and precious metals as people sought some haven for their money during the period of rapid inflation. Real industrial investment had virtually ceased in the United States and it was beginning to show. When it came to the competitive production of industrial goods, Americans became cognizant that for the first time since World War I, they were no longer the dominant industrial power. In industry after industry, the market for American-made goods shrank in the face of foreign competition. The main culprit was Japan.

Japan had many hurdles on the road to industrial prominence. Lacking virtually all raw materials necessary for industrialization and a destroyed industrial base, the Japanese faced problems so serious that a competition with a resource-rich nation such as the United States, should have been lost by default. Such was not to be the case, for Japan had several advantages that proved crucial. Although the Japanese faced the daunting task of post-war reconstruction, a task that had the profound effect of focusing their energies, the United States had grown arrogant and lazy—as the rich often do. The first industrial products from post-war Japan were cheap and crude, which reinforced American complacency. The Japanese became better very quickly. They had a host of well-trained and dedicated engineers who were not diverted to the arms business as is true in the United States

If the goal is to develop high-quality goods for the popular marketplace, civilian development is roughly 100 time as efficient as military-based development. By concentrating all their energies on civilian projects, the Japanese had found the ultimate industrial shortcut. Their first forays into the market were based on products with low costs. Before long, the Japanese had added world-class quality in design and execution. They mixed leading-edge quality with low price and the results were a dynamite combination in the marketplace.

Ideologically, the supply-siders were incapable of addressing the question of what form new investment should take. When cornered on the subject, which was not very often, the pat response was that the market would take care of those decisions. The supply-siders had a notion about who was best prepared to make those investment decisions. They believed that the rich would make the correct investment decisions. After all, the rich were rich because they had made correct investment decisions in the past. The basic supply-side notion was that if tax incentives were made large enough to interest the rich in productive investments, they would get out of their speculative activities and rush forward with a wave of investment that would restore America to its rightful place at the head of the industrialized world.

There was a huge flaw in this reasoning: most of the rich are rich because of predatory activities. Very few are rich because of constructive activities and most of this minority were poor when they started. For investment to produce industrial growth instead of more economic distortions, the investment must find its way into the hands of the industrial types—not the predators.

One of the pitches used to sell the supply-side agenda was the phrase "a rising tide lifts all boats." As used in this debate, this phrase was used to justify giving additional income to the rich because the ensuing prosperity would eventually benefit everyone. Prosperous rich people would spend money, which would create employment down the line. In days gone by, this concept was called the trickle-down theory. In reality, trickle-down is not an example of a rising tide lifting all boats but rather a hope that a *rising boat will lift all tides*.

Since wealth and power gravitate from the poor to the rich, raising the standard of living of the lower economic classes has eventually been beneficial to the rich. The reverse has not only never happened, but is extremely unlikely given the dynamics of human societies. Supply-side economics was fatally flawed because it was based on absurd assumptions. The supply-siders spoke for the interests of the rich who would be richer. They had no plan for improving the general economy nor did they worry about industrial development.

When the question of industrial investment was finally asked, the answer was to revert to plan A. The United States was still predominant in the manufacture of weapons. Warfare had been good for the economy in the past and in spite of the fact there was no one who really wanted to get into the kind of war with the United States that the United States wanted to fight, there was a way to have a war without the inconvenience of having a lot of funerals: get serious about the arms race. Ronald Reagan had been elected on a wave of discontent brought on by the perception that the United States was slipping from its predominant position on the world stage. Reagan argued that the problem was that America was insufficiently armed or belligerent. He called for the rearming of America.

It never entered his head that the problem was not insufficient spending for weapons but excessive spending. Reagan's World War II experience was a combination of the domestic excitement of home-front support and a decidedly non-lethal existence in Hollywood. For a large portion of his adult life, he had served as a corporate spokesman for the large defense contractor General Electric. Whatever the notion of a strong defense meant to Reagan, it meant fat order books filled with cost-plus contracts for his former employer. With the help of a like-minded Congress, most of whom had defense industries in their districts, Reagan pushed through legislation beginning the biggest peacetime arms buildup in United States history. The United States was about to embark on a peacetime war.

Not all debt creates the same amount of economic activity. If debt is incurred and the money is passed to the lowest levels of the economic scale, a welfare mother for example, the money makes many stops on the way to the pockets of the rich. Send money to a welfare mother and before the day is out, she will have passed some of it on to the grocer, who passes it on to the wholesaler, the processor, the farmer, the implement dealer, the factory worker, the factory, the steelmaker, the ore haulers, the iron miners, the coal miners etc. She will also spend money on rent which winds up in the hands of the rich the same day. One dollar takes a very long trip to the bank—the other, a very short one. Part of the economic stimulus created by welfare spending causes a great deal of further economic activity; part creates no stimulus at all. When the Reagan administration began to incur federal debt in the range of $200 billion per year, his bow to fiscal responsibility was to cut back on social spending, the route known to have a maximum economic stimulus for each dollar of debt incurred. He replaced it with militarism and usury, the shortest trips from bank to the bank (where money does no good) with the least amount of general stimulus involved. Construction, another maximum effect stimulus, had died because of the onset of monetarism in 1979. In the rush toward military spending, the construction industry was virtually left out except as a spin-off of the military build-up.

That the military should have gotten the lion's share of the limited industrial spending by the Reagan administration should not be surprising. What is important to remember is that Reagan was merely a salesman. It is very difficult to imagine any great military mind in a leadership position when it comes to the industrial aspects of warfare. Douglas MacArthur, clearly the greatest of the American generals of history, who also understood the importance of industrialization, would have been at a loss to describe the manufacturing process of most simple items in his armory. If MacArthur did not understand the nature of his industrial base except in theoretical terms, Reagan, a simple actor who had problems remembering the names of his cabinet members, was not even in the ball game. How wars are fought is a function of the arms. How and what arms are made is, and has always been, an industrial matter.

Where the money was to be spent was up to the arms makers. The arms makers are very careful to state that they are only responding to the wishes of their governments. This is the lie that arms makers have been telling since there were arms makers. The warriors are incapable of conceptualizing new weapons: they have great difficulties understanding the weapons they already have. The relationship between the military and the weapons makers can best be understood in terms of the relationship between a car builder and a race driver, or an airplane builder and a test pilot. In none of these cases can it be argued that the pilot, driver, or warrior is in the dominant position though they get all the attention. In the case of the arms builders, the military is a good customer that generally buys what they make at a price that makes the effort worthwhile. Some arms makers have no other customers.

By 1982, monetarism, and its most ugly manifestation—high interest rates, had bankrupted all the little borrowers. Banks smelled trouble—bankrupt borrowers lead to bankrupt lenders. International banking came very near to a total collapse when Mexico, an oil-producing nation, nearly defaulted on $100 billion in debt. The little borrowers had become very large. The bankers needed a big borrower with a big line of credit foolish enough to pay the prevailing interest rates. [2] The United States government would fulfill this need by financing a peacetime war on credit.

A tax cut to spur investment combined with wasteful program of increased military spending had a very predictable effect: it increased the debt of the United States government. Though they thought of themselves as the antithesis of Keynes and refused to mention his doctrines, the supply-sider strategy that was used is old-fashioned military Keynesianism. Lord Keynes taught that when the economic cycle turns down, the government should step in and prime the pump with deficit spending. As the money works it way through the economy, it creates more economic activity and the recession turns into an expansion. Jobs are created along the way which makes for new taxpayers.

Keynesian teachings represent a monetary compromise. Galbraith points out that Keynes' theories of monetary expansion leading to industrial growth were derived from the old Greenback theories. Keynes would put an English patina on the theory of monetary growth that would lend status to the basic ideas. But what made Keynes theories palatable to those concerned with money rests in the fact that he would prescribe that the money supply increase through borrowing. The money supply can be increased by the Federal Reserve whether it issues debt or not. The experience of World War II taught that borrowing is not always necessary.

When the United States dramatically increased its budget deficit in the early 1980s, bondholders the world over held their breath to see how much of the deficit would be paid with a simple increase in the money supply and how much would be funded by debt. Volcker decided by 1983 that it *all* would be debt. [3] Monetarism was granted a reprieve—the lenders of the world would not fail because of the default of smaller borrower. [4] Volcker got his reward—he was reappointed for a second term as chairman of the Fed.

Military Keynesianism on the Ground

BOEING IS AN EXTREMELY PROUD company that makes very high quality products. While Boeing's defense contracts are substantial, it has never neglected the civilian market and unlike other large aerospace contractors, Boeing make substantial profits from civil aviation. Virtually anything good that can be said about how a company is run can be said about Boeing. Other companies make jet aircraft but Boeing is responsible for the Jet Age. Selling aircraft to airlines, the job Boeing does best, is a business subject to external cycles. When the business climate is good, people do more flying and airlines buy planes. The energy crises of 1973 sent tremors through the airline industry. Jets use lots of jet fuel and the cost more than tripled. To maintain profits, airlines had to hike their rates, hold the lines on salaries, and cut back on new airplane acquisitions.

Like a responsible company, Boeing went back to the drawing boards and redesigned a new fleet of planes that were about three times as energy efficient. They also needed one less crew member; which addressed another concern of the airlines. New problems for the airlines: new solutions from the folks who build airplanes. Should have worked, but it did not. Interest rates had started to climb and the airlines found it more economical to pay higher fuel bills than it was to pay the interest on the borrowed money for a new fleet of airplanes. They kept the old planes in the air, a possibility given to the airlines because Boeing made such good airplanes in the first place.

Though Boeing had vanquished all its domestic competition—airplanes made by McDonald-Douglas and Lockheed were rapidly losing market share, a newcomer on the block was causing serious problems. The A300 Airbus, built in Europe as a government backed consortium between nations, was a very good airplane; in the same class as anything from Boeing. The Airbus had a tremendous advantage in sales over Boeing, however, for one simple reason. The Airbus, backed by governments who had a stake in the employment of Airbus workers, offered financing significantly below market rates—their method of combating monetarism. At the lower rates, the fuel savings suddenly made economic sense. The 757 and 767 series airliners are flying but in nowhere near the numbers had Boeing had the financing advantage of Airbus.

The loss of sales to Airbus was merely another setback to Boeing and Seattle, the community most affected. It is possible that Boeing would be happy and content to spend the rest of its industrial existence designing, building, and selling airplanes to airlines. The quality of the effort put into their airliners is sufficient reason for this belief. Boeing lost many military contracts because designing reliable and safe airliner is a very different challenge from building a bidding cost competitive military aircraft that is supposed to be slightly dangerous. The quadruple backup systems for flight controls of the 747 exceed what any military mind would deem reasonable. Boeing is so cautious that they are referred to as a belt-and-suspenders outfit within the industry (named after the fellow who is really worried about his pants falling down.)

By 1979 Boeing was in a state of virtual shutdown and Seattle was suffering very high rates of unemployment. Senior engineers, who had graduated with the highest marks in the hardest subjects humans have ever concocted for themselves to study, were standing in the unemployment line. This sight eliminated any purist thoughts about civilian production in the minds of Boeing management. What Boeing wants, Seattle wants, and so do their representatives in Washington. Boeing wanted orders and if the civilian sector was not forthcoming, the military would do just fine.

It is not that Boeing lacked prior contacts with military production. Boeing built some very important military aircraft in history including the B-17, the B-29, and the B-52, and Boeing had tried and succeeded in converting at least some of its expertise to productive ends. A 747 employs people for years after it is built, moving people and goods all over the world. The same cannot be said for an equivalent bomber. Part, if not all, of Boeing's heart was in the right place, which is more than can be said for those defense contractors who have no civilian expertise or customers.

Boeing went after the contracts for the B-1 bomber, a weapon that had been scuttled by Jimmy Carter for having no useful purpose as a weapon. Be that as it may, the B-1 was a very interesting aircraft that used much of the Boeing SST expertise gained before that project was scrapped. Boeing could at least profit from some of this know-how that would become valuable if the B-1 were put into production. Moreover, the B-1 was to replace some of the aging B-52 fleet. Even though North American Rockwell would be the prime contractor, Boeing looked on some of that business as theirs. Boeing wanted the B-1.

Jimmy Carter was correct about the B-1. It has no use as a weapon in the age of missiles. A change in payload would alter all that. The change was the air-launched cruise missile that Boeing would build. The B-52 could also launch the cruise missile but that was besides the point in the arguments over whether the B-1 should be funded. Congress was in the mood to spend money for weapons and the B-1 with the cruise missile seemed as good a place as any to spend it. The B-52 was 30 years old and that was reason enough. The B-1 had a powerful constituency. Rockwell, the prime contractor for the B-1, had managed to spread its subcontracting work over 426 of the 435 congressional districts. Even if the B-1 is an anachronism and a joke as a weapon, the cruise missile, which was used to justify its production clearly is no laughing matter. That Boeing is the prime contractor for the air-launched cruise missile is reason enough to worry about it as a weapon. Boeing makes things that work.

The cruise missile is really a very small, low-flying aircraft. It has no pilot. Instead it has a computer guidance system that, in theory, can literally guide it down the chimney of the target. It cannot be detected by radar because it flies beneath radar's clear range. It can carry a nuclear warhead. It is a very dangerous weapon. What started out as an employment program for the educated of Seattle became the hottest new issue of the arms race. It did not create all that much employment but it certainly made the world a much more frightening place to live. It also woke up the moribund peace movement, at least in Europe.

The peace movement of the 1980s made the cruise missile its cause celebré. The cruise missile, it was argued, is a first-strike weapon. Its phenomenal accuracy, at least as advertised, makes the cruise the perfect weapon for striking at the missile sites of the enemy. The ability to knock out the weapons systems of the enemy is supposed to mean that it would be possible to win a nuclear war because the enemy would have nothing to fire back. By raising such a fear, the peace movement actually helped sell the weapons in Congress. To the belligerent, the possibility of having a first-strike weapon was intrinsically good. To the less warlike who did not believe the cruise missile was a first-strike weapon, mostly because they could not imagine the United States ever becoming involved with such a weapon system, the protest against the cruise missile was merely evidence that those sneaky Russians were spreading vicious lies to stop the system. If the Russians were concerned enough about this system to stoop to lying about it to the peace movement, it must have some value—if only as an arms control bargaining chip.

Selling nuclear weapons to the United States is quite a feat. One weapon dropped on an American city would cause more damage than any man-made catastrophe in the history of the country. One hundred nuclear weapons could destroy all the major cities in the United States, permanently damage the ecosystems of the country to the point where food production was virtually impossible, and damage the infrastructure to the point where the survivors would be left with the industrial capabilities of the seventeenth century.

Since not many people, even among those so-called survivalists, have the ability to survive very long without their industrial umbilical cord, the number who would survive a small nuclear exchange of 100 weapons is very small. By the time 500 weapons would be exploded, more explosions would only have the effect of rearranging the rubble. There are already 50,000 nuclear weapons in the world. Anyone who could sell the governments of the world some more of these useless devices is either the world's greatest salesman or has the world's dumbest clients. In the case of Ronald Reagan, both were the case. Worse, as the world's citizens finally awake to the idea that modern industrial war is impossible, they face the nearly impossible task of cleaning up after the arms race, burdened by debt.

THE SELLING OF CATASTROPHE

SELLING NUCLEAR WEAPONS to a nation that already has far too many would seem a daunting task. It really was surprisingly easy. At the head of the cast of characters was the President of the United States who had honed his selling instincts for years pushing light bulbs and laundry soap. Good selling is probably instinctive and those who can sell, can sell anything. No question about it, Ronald Reagan was a very gifted pitchman.

To his advantage, he entered office with the country in economic and social distress. No one could seem to put a finger on what was wrong but everyone knew that life had gotten a great deal more dreary since the optimistic days of the 1960s. The notion that the United States was a target for international humiliation was widespread as a result of Iran, the fall of the Shah (an American puppet and arms customer extraordinaire), and the seizure of the American embassy staff. This final outrage was particularly egregious because it was an attack on official Washington. Under such circumstances, Reagan's tough talk had the maximum effect.

The selling of an arms race in peacetime was couched in phrases that recalled a host of questionable assumptions. Investing in waste as a cure for economic problems was merely the final absurdity. Warfare has been profitable in the past, but only for the winning side. Even winning does not guarantee economic gain but losing certainly insures a loss. Arming a nation so warfare is impossible, the stated reason for Reagan's arms buildup, is a certain economic loser. If weapons can be defined as tools to seize the wealth of someone else, then nuclear weapons are not, strictly speaking, even weapons. Nuclear weapons can only destroy the assets of an opponent. By their very definition, nuclear weapons make warfare obsolete as a form of organized armed robbery. Unless they are used, nuclear weapons only destroy the assets of the country that possesses them. The ultimate peculiarity of nuclear weapons manufacture is that such weapons are extremely destructive, especially in economic and environmental terms, to the nations which make them. That a wholesale nuclear exchange would destroy the biosphere, there is no doubt. It should be equally evident that the mere manufacture and possession of nuclear weapons is extremely destructive too.

Objective thinking about nuclear weapons was beyond the scope of debate over military spending. Americans were told that the Russians were spending a greater percentage of their gross national product (GNP) on militarization. These figures were heavily doctored. The most dishonest example was figuring the costs of the Soviet armed forces as if they were paid amounts equivalent to the volunteer members of the American armed forces. Even if the figures were true (which was not often), the question should have been raised: "Do the Western Democracies want their economies to be as thoroughly distorted by militarism as is the economy of the Soviet Union?" "The cost of defending the nation is high," it is argued on the other side. "Whatever the Soviet Union is doing, Americans must match."

That this is a distorted argument is demonstrated by the fact that the United States led every round of the arms race. Given the nature of the Soviet system, it was impossible for them technologically to do anything but try to match the west. What the Soviets learned from World War II was that since they are unable to match the other industrial nations in technical expertise, their response will be to make very many simple, but effective weapons—a strategy that is best characterized by "make it cheap, make it work, and make a lot of them." The Soviet economy, distorted as it was by arms manufacture, was hardly a model to emulate—especially since their arms, in general, were not very good. When it came to the Soviet nuclear arsenal, the weapons, though primitive, were clearly good enough to do a tremendous amount of damage. True to form, the Soviets had a great many of them.

The concept of nuclear deterrence is based on the notion that in case of attack, enough weapons would be fired in return to cause such great damage, that no one would risk attack. What constitutes a sufficient deterrent? The question cannot be raised because the answer is so absurd. When each side has over 20,000 warheads, the main headache for military planners is finding sufficient targets worth destroying. An American nuclear planner, when pressed to name the lowest priority target in the Soviet Union, told his audience that it was an open field outside Moscow that could be used as an emergency landing strip by returning Soviet bombers. The fascination with counting weapons on both sides indicates the gross levels of ignorance of what an atomic weapon can do.

If even a small fraction (such as 1/5) of the atomic arsenals of either side were fired at the other, even if there was no retaliation, the biosphere would suffer such damage from radioactive poisoning, ozone depletion, and the cooling effects of smoke that it is very unlikely that the planet could support human life. Nuclear weapons are not like arrows, spears, bullets, or howitzer shells. After 200 exploded on the entire planet, any more are totally irrelevant. Being able to destroy the biosphere 31 rather than 29 times has absolutely no meaning. Anyone who thinks it has meaning would probably think that singing 36 verses of "On Top of Old Smoky" was three times as much fun as singing 12.

The most ridiculous discussion concerning nuclear weapons deals with the subject of accuracy. Accuracy has been the goal of weapons makers since the dawn of recorded history. A great archer or rifleman was considered great because he could hit his target. With nuclear weapons, hitting the right hemisphere is close enough, especially if enough are exploded. Currently, missile accuracy can be measured in a few hundred meters—a great deal more accurate than the necessity of hitting the aimed-for hemisphere of the planet. Spending money to make missiles more accurate is absurd but Reagan was able to convince legislators that such expenditures were necessary.

That nuclear weapons were sold in the terms of nineteenth century militarism would have been extremely funny if it were not so tragic. Showing resolve became fashionable. "Would the Soviets think we were weak?" the question was asked, "if this country failed to appropriate enough for its defense?" The hallmark of this epoch was that politicians of all persuasions were much more afraid of being weak than of being absurd. In the stampede to prove toughness, none of the tough questions was asked: Does more money for weapons make for a more effective defense? Does military spending harm the economy, especially if done with borrowed money? Does military spending harm industrial competitiveness? Are other measures of strength more important than military might and are these forms of strength harmed by increased military spending? By 1982, it was clear that the combination of monetarism, high interest rates, the tax givebacks to the rich, the resulting high American government deficits, and the industrial planning schemes of the Pentagon were a certain prescription for economic catastrophe. The great riverboat gamble had failed. The American economy began to unwind at a pace that stunned the experts.

The industrial base, except that which was tied to the manufacture of weapons, became mostly uncompetitive. The huge borrowing needs of the American government kept real interest rates high enough so that it attracted foreign capital. The massive exchange of foreign currencies to dollars as foreign investors sought to avail themselves of American opportunities had the side effect of driving up the dollar in relationship to those currencies. The overvalued dollar completed the destruction of the industrial base—imports into the United States became cheap, export growth became nearly impossible. Export industries shut down so that a falling dollar in the future would have little effect. The methods for restoring the balance of trade had been ruined.

The United States which had a total trade deficit for the whole of the 1970s of $20 billion, ran a trade deficit of over $100 billion for 1984 alone. Between 1982 and 1985, the United States slipped from being the world's largest creditor nation to the world's largest debtor nation. By the fall of 1985, American agriculture, the single most productive industry in the history of the planet, was on the verge on complete default of over $220 billion in debt, a figure large enough to certainly bring down a host of lenders. Industry after industry, from shoes to textiles, farm machinery to computers, and steel to consumer electronics was reeling.

No realization of this ever reached Washington where the politicians, their servant-bureaucrats, and the scribe-journalists touted the amazing recovery of the economy. The recovery they talked about was a matter of Americans doing nothing but buying undervalued foreign merchandise on borrowed money. The American industrial plant was operating on nothing but inertia.

As the 1980s drew to a close, military competition became obviously futile to even the most rabid militarist. Military Keynesianism, the engine of American industrial growth for 45 years, has lost its luster, yet the United States keeps hoping and praying for one more military-driven economic salvation while the banking structures come crashing down—proving yet again that when the real economy has been damaged long enough, even the bankers find out about it.

Did the bankers and politicians and arms makers set out deliberately to destroy the industrial base of the country? It is unlikely. Even those who would posit the notion that Reaganomics was good for the rich fail to understand that with the decline of the industrial base, the riches go as well. The rich may have thought they were getting richer but this is only a case of delusion. Mistaking money for wealth, money for capital, business for industrialization, and weapons for strength led to the immiseration of the poor and impoverishment of the rich. Borrowing had only created debt, not investment. A failure of understanding had created a failure of economics of tragic proportions. There were no winners from the great riverboat gamble.

NOTES ON CHAPTER NINE

[1] Seymour Melman *Profits Without Production,* Knopf (1983).
The shocking waste of industrial talent in the manufacture of arms is best recorded by Melman. It is possible that people who have worked in "defense" contracting long enough have formed permanent, bad habits as Melman argues. I maintain that a builder is a builder and almost nothing can drive it out of a person. There may be mediocrities in "defense" industries because they can afford them. Those people would not make much difference under any circumstances. If real builders are given real assignments, they seem to find a way to solve the problems. The problem with weapons development in peacetime is that work tends to become meaningless. How excited can a person get, after all, about the next round of weapons if the last round has never been tried. Under such conditions, the good builders become positively Baroque. Laser-guided tank aiming devices are merely the twentieth century equivalent of the Cathedral Gargoyles of the twelfth. The same kind of mind is at work.

[2] Penny Lernoux *In Banks We Trust,* Anchor Press (1984).

Bankers choose to portray themselves as sober, civic-minded citizens who provide safe haven for people's money while providing a source of capital for investment in the growth and improvement of society. Of course, some bankers are exactly like that. *In Banks We Trust* is about the rest of them.

Banking is by its nature the quintessential, non-military form of predation. However, when banking leads to the general enrichment of the society, no one minds or should mind that bankers are enriched in the process because they have provided a valuable service. Since the key to any productive investment is

NOTES ON CHAPTER NINE

patience, bankers sometimes are the best source of capital because they are most patient. The case of Toyo Kogyo of Japan, makers of Mazda cars, is illustrative of patient banker and their ultimate reward. Mazda bet heavily on the rotary engine. The rotary engine had many development problems—especially durability, emissions, and fuel economy. Just when the problems of emissions and reliability were solved, O.P.E.C. crises #1 collapsed their hard-won market. Through all this time, Toyo Kogyo lost money. But their backers did not lose faith and Mazda responded to their problems by turning out a series of winners that catapulted them into the fourth place among Japanese manufacturers. As a former owner of a Mazda RX-7, I say with reason that the decision to stay with Mazda was one of the wisest banking decisions ever. Some German and Scandinavian bankers are much the same as their Japanese industrial counterparts.

The temptation for bankers to let predation get out of hand usually proves irresistible, especially in urbanized societies where the connection between the figures on the computer screen and the activity behind it is almost impossible to see—even for those who would wish to see it. *In Banks We Trust* covers the devastation wrought by these bankers who are either grossly ignorant or have real criminal intent. Since bankers portray themselves as highly competent, the temptation is to brand all their activities as criminal. Penny Lernoux occasionally falls into this trap but she seems to have extensive justification for believing much of banking has, in fact, become the province of persons who extend their criminal behavior far beyond simple usury. The people and practices she indicts should be rooted out—especially by those honest bankers who actually contribute to growth. Banking of some sort is necessary, what Lernoux describes is not.

3 William C. Melton, *Inside the Fed; making monetary policy*, Dow Jones Press (1984).

This book is a relentlessly conventional description of the workings of the Federal Reserve System. However, it contains one giant gem of information—it describes the Volcker decision that all government borrowing would result in increased debt. This information has been hard to confirm elsewhere but as Melton is a former Fed economist who writes as a cheerleader for the wisdom, power, and "independence" of the Fed, there is no reason to believe that he is not telling the truth when he describes this act of high treason.

4 Mark Singer, *Funny Money*, Knopf (1985).

Singer's book misses the point as to what went wrong with the Penn Square Bank of Oklahoma City. Yet, the book is worth reading for its description of how far bankers will go to ignore the fact that returning 17 percent interest means more than shuffling papers or electrons. The whole economies of Texas, Oklahoma, and Louisiana went into a state of virtual collapse because of the same factors that did in Penn Square. It will be difficult to write off all of this to "colorful" bankers and their reckless habits as Singer does in his book.

NOTES ON CHAPTER NINE

The most interesting fact about the collapse of Penn Square it that it caused a crises at Continental Bank of Illinois which led to the first big bank bailout of the 1980s. This is important because it established the "too big to fail" philosophy which said that when a bank reaches a certain size, the various agencies of government will prop it up no matter how stupid or criminal were the actions which got the bank in trouble in the first place. With this assurance, banks would become more reckless and criminal until the whole rotten mess would be exposed in 1990 and taxpayers would be forced to bail out the banking system as a whole. *Funny Money* should be read because it traces the roots of the banking crises of the 1990s.

PART THREE:

THE INDUSTRIAL ENVIRONMENTAL SOLUTION

CHAPTER TEN

DO PRODUCERS HAVE A PLAN?

The best way to predict the future is to go ahead and invent it!
Anon. (Nominated for "Official Producer Slogan")

WHEN THE ECONOMY TURNS SOUR, as it did in earnest at the end of the 1980s, concern for nature and the environment evaporates like the morning dew. Far from having lasting impact, Earth Day 1990 faded from public consciousness faster than a bad advertising campaign. Compared to the problems of daily survival in times of economic hardship, environmental activism seems like a hobby for the idle rich.

If environmental action follows the strategy of the old conservationists, modern industrial economies and their workers are considered targets. Industrial activity, according to these first-generation environmentalists, is the problem.

Conservationists believe nature must be preserved. Royal hunting estates became the model for a more democratic American incarnation: national parks. If American wetlands are saved, it will be thanks to the duck hunters. The extremely important work of saving species biodiversity is the shining star of the conservation movement.

This is still a strategy of the hunting rich. Nature is more important than people—the spotted owl is more important than the jobs and futures of timbermen. This produces genuine class bitterness. The rich, who want the parks of old growth timber preserved, forget that the houses they own were built with lumber from such forests. Likely, the rich hunter has taken far more than a fair share of the resources of the earth. To the woodcutter, the lifestyles of the urban rich will kill those forests with acid rain anyway, so why not cut down trees while they may still be used for lumber?

Conservation stands charged with elitism. It is guilty as charged! Conservation techniques deserve the mantle of elitism because it is an elite idea—in all the varied meanings of this word; however, so long as conservationists understand that their strategies will solve but a vital fraction of the environmental problems—they are but niche players, their status is assured. If the rich do nothing in their lives but preserve biodiversity, they will have justified their very existence. Yet it is essential that conservationists support an industrial environmental strategy or their very best efforts will fail.

Conservationism has its excessive manifestations. There is a "deep ecology" movement in Scandinavia that proposes to limit production practices to those in existence before the nineteenth century. As the planet could only support about 15,000,000 people using these methods, deep ecology fails as a philosophy because it does not discuss what is to happen to the rest of us. As Europeans are understandably sensitive about any plans that involve large population reductions—having heard enough of those sorts of plans for one century, deep ecology unfortunately has been tagged with the label "Eco-Fascism" —unfortunate because deep ecologists have much to contribute to environmental thinking.

In the division of labor necessary for an environmental solution, conservationists must understand that while they maintain and expand the nature preserves, educate the young, and collect specimens so as to preserve biodiversity, they must support the human needs of production. Anyone who has heard trade unionists and environmentalists shout at one another understands the need for a truce in this class warfare. If production promises to support the agenda of conservation, conservation must promise to eliminate the economic impediments to the production and design of an environmentally correct new industrial infrastructure. The truce must be permanent because this job will be large, time-consuming, and very expensive.

With truce in hand, the attention shifts to the producers and the question, is a new industrial order possible? The question is a valid one. Fordism is the economics of production, and Fordism's internal contradiction is environmental destruction. Leave aside the philosophy of Fordism and concentrate on Fordism's model product—the automobile. The fossil-fuel burning car and the urban sprawl it made possible is the single largest extant environmental problem.

In Europe, where the concept of Fordism has a specific meaning, the early results of an industrial-environmental truce are in, and the signs are positive. The modifications to Fordist philosophy are collectively termed post-Fordism. In its primitive forms, post-Fordism is strictly a production philosophy. Industrial design is still absorbing the radical new possibilities and power available to them through the marriage of machine tools and the computer. Tools are ahead of imagination in this phase of thinking, but imagination is catching up.

Because industry is so important to German existence and self-definition, the Germans never considered deindustrialization as a solution to environmental problems. *Die Grünen*, which shares most political assumptions of the Social Democrats (SPD), was forced to develop an environmental strategy compatible with the SPD economic principle number one—workers must prosper, therefore industry must prosper. If post-Fordism has a political dimension, it is the early industrial-environmental philosophy forged in the Green-SPD compromise.

The Germans have a history of this sort of social contract. When Martin Luther was drawing up the doctrines of his new creed, he specified that each good Lutheran home have a Bible and hymnal. Early German printers were important to Luther—in many ways they saved his life and started his movement. The economics of Luther was simple—stop sending money to Rome to build big churches and spend it at home on the products of German printers. Mandatory literacy became an article of faith. German printing prospered. A technological improvement gave distinction to the new social order.

With today's industrial reformation, the sequence is similar. The new technological capability came first and then the question was asked, "now that it is possible to produce anything conceivable in any quantity, is it possible to produce a modern society without destroying nature?" No one knows the answer to this question but simply contemplating the possible answers has producers muttering happy phrases—hundred-year projects—the pyramids will be forgotten—watch my motion.

Producer enthusiasm for this post-Fordist industrial truce is tempered by a realization that their human needs may not be considered. Producers know that building the pyramids was no day at the beach. They do not want to be slaves. Post-Fordism is fine if it retains a Fordist paycheck.

Environmental Fordism requires a change in social and economic assumptions. Building something better than a Fordist technological infrastructure will be a difficult task. Producers cannot afford to be distracted by unemployment or failing businesses while inventing the sophisticated pieces of the new industrial order.

The timing could not be better, for a transformation to post-Fordist economics, than the 1990s. All rational minds agree that current industrial practices are environmentally unsustainable. Many believe that a new industrial order is possible—that the tools have already been invented. The end to the cold war has caused thousands of highly skilled producer specialists to lose their jobs. The economics invented to justify the cold war is in disarray—no possible good can come from flogging that dead horse again! New tools must change the old economic assumptions. The time is now!

THE PROMISE

ALMOST ANY SERIOUS OBSERVER of world events knows that something is clearly amiss. The economic and monetary problems reflect the industrial crisis. Even though the monetary problems have as their root causes the industrial crisis, wrong economic decisions have made the industrial crisis infinitely worse. It might even be argued that the economic policies caused the industrial crisis, although that would be giving economics more credit than it is due.

At the root of the industrial crises is the worst of all possible assumptions: that geometric growth rates in any endeavor can be maintained in a finite ecosphere. This assumption is in error although most people act as if it were not. The language of geometric growth surrounds the discussions of business, finance, and government. Any time growth rates are expressed in percentages such as for a population growth of two percent per year or an accrued interest rate of twelve percent per year, the intrinsic assumption of geometric growth is built in.

Geometric growth presents two important challenges. First, any geometric curve eventually reaches the stage where the growth line is essentially vertical which, for all intents and purposes, means that the rate of growth is infinite. Second, the area under a geometric curve for the final doubling time equals all the other areas under the curve combined. Problem one leads to collapse for the simple reason that infinite growth of anything in a finite biosphere is impossible. As growth rates approach infinity, a collapse, usually catastrophic, is inevitable. Problem two explains how soon the catastrophic collapse will occur as it gives a picture of how fast a finite resource is being used up.

This industrial limit will be faced in one way or another. The choice is between a valid set of assumptions or the four horsemen of the apocalypse. In an economy, such as the United States', assumptions of geometric economic growth are built into a host of forecasts from budget projections to future Social Security payments. But economic growth must be based on some form of meaningful activity. Economic growth means more than bigger numbers; it means greater outputs of tangible goods and services. Economic growth is founded on industrial growth. Without growth in the industrial sector, larger numbers are merely a form of papering over the problem.

Industrialization, in its current realization, cannot grow at geometric rates. In many areas, industrialization has already reached its resource limitations. Domestic United States oil production peaked in 1972. Geometric growth rates in oil consumption would only exacerbate the balance of trade problems. Forms of agricultural production are unsustainable. Row crops, such as soybeans and corn, consume, in the form of erosion, about a ton of fertile topsoil for each ton of crop produced. The resource of topsoil can be renewed but never at such rates.

Industrialization destroys resources by rendering them unfit for human consumption. Water is plentiful, but fresh water is not plentiful, comprising only about one percent of the total, and water that is healthful for humans is a fraction of that one percent and shrinking daily because of pollution. If growth means pollution, then growth will soon encounter the limitation of a finite supply of clean air and water. As the industrial system is currently organized, not only is further growth virtually impossible, but also creates the danger of catastrophic collapse. Flogging the current system into further growth, even at past historic rates, is an exercise in futility. Business as usual is clearly out of the question.

One of the most common variants of the business as usual solutions to the industrial crises is a future of high-technology and information. As the industrial system grows more troubled, high-technology solutions become increasingly discussed. There are many problems with the high-technology solution beginning with a working definition of high-technology; it means almost anything. It could be defined as anything that is on the leading edge of human knowledge. If that is true, then high-technology as a solution for the future encounters the reality that all the problems are not yet known.

Early discussions defined high-technology as high profit, young enterprises such as the computer industry was from 1950 to 1980. This definition is appealing because there are very successful models to emulate. In 1982 when the great industrial giants such as automobile and farm equipment makers were staggering under the policies of monetarism, a group of young Senators and Congressmen in Washington took to calling themselves "Atari Democrats." It was a simple matter of the aging process of industrialization, they reasoned: the new computer-based industries would replace the old smokestack industries. No sooner had the press latched on to the name of these "forward-thinking" representatives, than the real Atari announced huge layoffs of domestic workers and a large-scale shift in employment to East Asia—so much for the Atari solution.

High-technology could also be described as any industry related to defense or medical supply. These are growth enterprises, to be sure, but they are not good examples of industrial growth. Unless the military is used to seize the wealth of others, investment in military high-technology is merely an exercise in waste. Medical high-technology, for whatever good it may do, is also from a purely social investment point of view, wasteful, because much of it merely prolongs the agony of dying.

Military and medical high-technology are enterprises entered into by societies that are already rich. These are not the means for getting rich in the first place. It is not meaningful to discuss solutions that replace the methods of becoming wealthy with enterprises of the already rich. If this is high-technology, then high-technology is no solution for the industrial crisis. Behind most talk about growth in high-technology are the same assumptions that have always misguided industrialization.

If high technology is a slippery concept to nail down, the promises of the information age are even more elusive. Computers are fascinating devices, filled with the promise of greater wisdom through greater information. Having a machine with perfect memory is very appealing to humans with imperfect memories. Connecting computers to tools may be the most significant industrial advance in the past fifty years.

While information and wisdom are clearly related, the possession of information guarantees nothing because perfect information does not mean that the right questions were asked. Computer buffs have given this phenomenon the name of GIGO, which means garbage in—garbage out. Faith in the powers of computers must not be overdone. While information is valuable in plant agronomy, architecture, and textile manufacture, information, by itself, never fed, housed, or clothed a single person, for work must still be done, decisions must be made, and assumptions constantly reviewed. This work must be done by humans. The increased information available to humanity, brought about by thirty years of computer manufacturing, has unfortunately done very little to increase human wisdom. Computers can help answer some of the great industrial questions but only when the real problems are addressed.

The information age is not an end in itself, but only a means to an end. Those who take comfort in the possible growth in information processing at the expense of the basic life-support function of industrialization have obviously mistaken means for ends. This confusion means large increases in the bureaucracy and overhead sectors of the economy will continue to be treated as economic growth, which in turn, masks the industrial decline and makes getting to the heart of the problem that much more difficult. As a result, while the promises of the high-technology and information scenarios seem to be real solutions for the future, they are really only mutations of the business-as-usual scheme.

Alternatives to the business-as-usual scheme have been proposed. If industrialization has caused so many problems and is unsustainable, one proposal is to SCRAP INDUSTRIALIZATION AND GO RURAL. This alternative goes by many names: intermediate technology, sustainable growth, the solar alternative, and the like. These alternate schemes are unlikely to solve much because it is impossible at this stage of human development to go backward to a more simple time and lifestyle. Anyone who believes that going back to an earlier stage of industrial development is the solution should, of course, be encouraged to try. Those who have tried are astonished at how difficult such a move is.

Industrialization is not the result of some well-orchestrated conspiracy by narrow vested interests who set out to create disaster, hardship, and misery. Most often, industrialization was the product of human efforts to make life easier. The simple life was very difficult. It still is.

The Mother Earth News, a monthly manifesto extolling the virtues of the rural existence, is illustrative of the problem. The key ingredients of the *Mother Earth* philosophy are a devotion to simple tools, simple foods, simple occupations, and simple dwellings. The *Mother Earth* philosophy is replete with contradictions. The magazine will write about homes built by slopping stucco between short logs stacked like cordwood. Not only is the stucco mix a product of advanced technology, which calls into question the purity of such a scheme, but the resulting home is an extravagant energy waster. An article appears that tells readers how to make lawn furniture out of poly-vinyl-chloride (PVC) tubing for fun and profit. PVC tubing is the product of very sophisticated and environmentally hazardous industrial processes. Building lawn furniture, or anything else, from PVC tubing, is a simple task, but hardly a back-to-nature enterprise. To complicate matters, selling PVC lawn furniture or organic honey, or any of the other *Mother Earth* suggestions, means that these plans are contingent upon someone else becoming a customer who has not dropped out of the industrialized, monetized economy.

The back-to-nature movement is a harmless diversion. It is an expensive hobby for the terminally nostalgic that runs aground on its own inconsistencies rather than a solution for the problems of industrialization. When even small disasters, such as an appendicitis attack, strike the practitioners of the primitive philosophy, purity flies out the window and the rush back to the twentieth century is on.

It would be easy to dismiss the back-to-nature crowd as a group of cranks, except for their fascination with solar power. According to them, solar power will solve everything from acid rain to excess concentrations of political power; and they are correct except for one glaring problem—harnessing solar energy does not involve primitive technology. Nor, for that matter, is it particularly cheap.

If high-technology can be described as a young technology that humans do not know very much about, and are in the process of understanding, then solar power is properly considered a high-technology endeavor, though it is rarely treated as such. Humans know a great deal more about powering their industrial societies with nuclear fission than with solar power. For all the real advantages of solar power, very few successful examples of its use have been built.

Amory Lovins, the author of an excellent book on the potential for solar power called *Soft Energy Paths*, built an experimental house in Colorado that would demonstrate, he claimed, that properly built, a house could become a net energy exporter. Lovins should be congratulated. Very few people have the courage to put their money where their beliefs lie. It should be noted, however, that Lovins's house is not the success expected. It cost more than $600,000 to build which means that it is hardly the kind of solution that has mass appeal.

Current photovoltaic (PV) cells, devices that convert sunlight directly into electricity, are still very expensive. Many people are seeking a way to make PV cells efficient from the standpoint of energy consumed in production and costs. PV cells in sunny areas hold great promise.

Even energy conservation, a most laudable goal of the back-to-nature crowd, is not a primitive technology proposition. The Swedes, easily the most advanced people in the area of energy-efficient housing, have found that making homes energy efficient involved such sophisticated technologies as CAD/CAM manufacturing, landscape influenced microclimates designed by computer modeling, microfine production tolerances, and custom extruded synthetic rubber seals.

Saving energy is not simple if a person wishes to remain warm in a cold climate. Quadruple-glazed windows are not simple to make or cheap. Wood stoves that are not grossly inefficient and polluting are expensive and sophisticated. In reality, primitive technologies and temperate climates are incompatible. Primitive or intermediate technologies may have a role in the industrial development of societies in warm climates, but as solutions for the industrial dilemmas of the already industrialized societies, their use is limited.

Another solution to the industrial crises could be called "spread the wealth around." It is argued, that all the real problems with industrialization have been solved. All the great discoveries are behind us and only the political considerations of distribution remain.

Humans have harnessed electricity; extracted, smelted, and fabricated virtually every metal on the periodic chart; made tools so sophisticated that further development is a waste of time; and made synthetic compounds to meet any known need or want. Instead of making a faster car, the argument continues, the time has come to make a more durable one, that could be used for transportation needs in developing areas of the world. The current level of technology is high enough: What purpose is served by going for a little final increment. In fact, real needs are being sacrificed to planned obsolescence.

Like most proposed solutions for the industrial question, the plan to spread the wealth around has been treated seriously because there is a very large element of truth in it. There is something a little silly in the amount of engineering devoted to making a Mercedes go 143 instead of 139 m.p.h.—especially those to be marketed in the United States where the speed limit is 65 m.p.h. There is some truth to the notion that the major discoveries in industrialization may have already taken place. It is also true that while the junk yards of industrialized nations fill up with cars that have died before they had to, major transportation needs in the underdeveloped world go unaddressed.

Unfortunately, the wealth-spreading solution is a utopian fantasy. The political and economic problems inherent in the proposals to spread the wealth are enough to scuttle them. Industrial nations have not been very generous in giving away their acquired industrial wealth and there is no reason to believe this situation is soon to change. This leaves trade as a distributor of wealth—another less than ideal model based on past experience. The nonindustrial nations have only natural resources and cheap labor to trade for industrial goods. Under current arrangements, trade only to makes debtor nations out of nonindustrial ones.

Even if the utopians could alter the hearts and minds of those who arrange foreign aid and trade so that the nonindustrialized nations could enjoy a better economic arrangement, the problems would not go away. The current realization of industrialization is a demonstrable failure in the industrialized nations. Exporting such industrialization means exporting the problems. Making vehicles durable and available to more locales worldwide may seem like a noble objective, but everyone must compete for the same source of fuel—and petroleum is a finite and diminishing resource. Virtually all industrial technologies require the consumption of fossil fuels. Making such technology available to more people does not solve anything—it makes matters worse.

Moreover, some technology is intrinsically dangerous. Exporting a chemical industry does not spread much wealth around. Ask the people of Bhopal, India. Spreading the wealth around is a wonderful idea. Unfortunately, the burdens of industrialization are not a real form of wealth. When the problems of industrialization are solved, spreading the knowledge around may be worthwhile. In the meantime, spreading the wealth is impossible because it is mainly the costs and burdens that are spread, seldom the benefits.

There is a real solution for the industrial crises that has been proposed—often called REDIRECTED RESEARCH. If research and development could be directed at the problems of industrialization instead of projects replete with waste, real progress could be made. Thousands of items from automobiles to weapons, from computers to airliners, are as good as they need to be. In five seconds a computer can process information that a human could not read in five years. Do computers really need to be faster? Airliners could be made faster, but what would be the point? They could be made larger, but what would that solve? Weapons are already beyond human comprehension and automobiles are at the limits of hedonism. Are there not better spots for industrial talent?

Redirected research has an enormous drawback: it is a solution that requires a widespread consensus as to the nature of industrial needs. It also requires direction and planning. Planning, the heart and soul of industrial enterprise, is treated as an evil. In the United States, planning is often seen as a "commie plot!" Central planning is seen as the very antithesis of democracy, the name given to United States industrial capitalism.

Interestingly enough, this irrational cry is often raised by the industrial class in pursuit of its own interests. Those who produce anything are in mortal terror that some bureaucratic clown, a known industrial illiterate, is going to make decisions for them. From farmers to scientists, they believe that though the government may be a suitable collector of funds, it is a very inept supervisor. "Look at what central planning has done to Russian agriculture, or science, or product distribution and quality," they say and all the nation nods its head. Moreover, the industrial class can make the point that directed research almost precludes the possibilities of accidental discovery. The whole chemical industry can trace its roots to the accidental discovery of an artificial dye found while the researcher was looking for a way to make synthetic quinine.

Accidental discoveries are rare, and the industrial types know it, so this is just another diversionary tactic to keep as much supervision as possible from their lives. They know that only about 5,000 scientists in the United States get to name their projects. Obviously, most research is *already* directed. Furthermore, accidental discoveries are usually related to the original focus of inquiry. The first artificial dye was found while a man was exploring in chemistry. Dye was not found as a spin-off of the attempts to fly. Directed research is a valid notion for one reason; answers are more likely to be found if people are looking for the answers. In truth, the arguments about directed research are merely arguments about who will do the directing.

If the argument over directed research is essentially a nonargument, the debate over industrial planning is similarly specious. The question is not whether the United States will or will not have industrial planning, but what form it will take. Currently, what industrial planning exists is done by the military. Current Department of Defense guidelines specify that production must be done in such a manner that at least one work shift is totally automated. The reason given for this requirement is, not surprisingly, national security. Even a treacherous strike by workers would not threaten supplies in time of war.

This argument is total nonsense but everyone has agreed to assert it with a straight face. If there ever should be an East-West conventional engagement, the destruction would last a few weeks, maybe a few hours before World War II would look like a hiccup. Further production, even at current rates, would not be enough to change the outcome of the battle. Of course, if the war should go nuclear, the ability of robots to make battleship shells will be irrelevant. The reason the military made automation a requirement is that the robotics industry needed research money and that was the easiest way to get it. The argument of national security is a very good sales close.

Industry uses the military shamelessly to pick up its bills by having the military require what industry can make to be made. The military is a very good customer in that way. If someone wants to produce something in the United States' commercial sector, life can be very difficult indeed. Product errors, even if accidental, can bring enough lawsuits to bankrupt the company. The military is not so picky. If a building contractor misguesses on the cost to build a home, he eats the loss. Not so with the military. The military knows it is buying products that are not fully developed and expects to pay the costs of development. Best of all, if the product fails to work, the military does not seem to mind. The representative of industry stands next to the general, who is supposed to see the project through to deployment to be promoted, watches the same missile miss the same target, pats the general on the shoulder and says, "But that's the way it's supposed to work." The general merely nods in agreement. After all, if it comes to that, will it really matter if 30 percent of the nuclear warheads fail to explode? Industry puts up with military direction because it is easily manipulated and comes with funds for development. If industry tries to sell civilian projects to the government, the projects are called pork barrel, perpetually fought over, and frequently underfunded. Being paid well and called a patriot is a better alternative for most people.

Redirecting research and the industrial path to which such research would lead, is really the only hope for solutions to the industrial crises. The dilemmas of redirected research, however, must be addressed and the most important one is that of direction. Fortunately, or unfortunately depending on how one views these things, the industrial crisis is so advanced that the direction of research and development is becoming obvious. The time has come to finish the industrial revolution and close the industrial loop.

CLOSING THE LOOP

IF INDUSTRIAL DESIGN has one great fault, it is linearity. Industrial processes are begun in the mines and end in the junkyard. This earth to earth movement encompasses a whole host of varied steps that have direction as its common denominator. A modern automobile contains as many as 20,000 individual parts. Each of these parts is the product of many separate manufacturing steps. To make a simple part like a piston, for instance, bauxite must be mined and processed into aluminum, the aluminum must be correctly alloyed and either cast or forged into an approximate shape, and the final shape determined by machine tools.

Pistons may be very simple but they have required the expertise of mining engineers, mining equipment designers and manufacturers, shippers and shipbuilders, metallurgists, smelters, machinists and machine tool designers, in addition to the automotive designers. According to Marx, all these people have been adding to the value of the original material of bauxite by adding their labor. When the piston has reached the end of its useful life and is discarded, the reality is that these people have been engaged in the process of converting precious natural resources into garbage. This is not to denigrate the piston while it is fulfilling its design function. The fault lies not with the piston or the people who made it, but with the one-way path to its final disposition.

Humans throw things away because they do not know what else to do with items that no longer fulfill their design function. Aluminum in the form of a worn-out piston has the advantage of being more valuable than the bauxite because the processing necessary for its conversion has already taken place.

Yet, worn-out pistons are not considered valuable because the resource is too scattered and almost inextricably entwined with other end products of technology. To mine aluminum as pistons requires that engines must be disassembled. Thousand of disassemblies would be required to obtain the aluminum contained in one big scoop of high-quality bauxite. Even the recovered pistons would not represent a supply of pure aluminum. Alloying, the process of introducing minute quantities of impurities to alter the working characteristics of the basic metal, is not easily reversible. It is much easier to introduce exact quantities of impurities than to remove them, especially if the nature and quantity of the alloying impurity is unknown.

The second law of thermodynamics states that energy always moves from a more organized to a less organized state. The flame from natural gas (methane) burns hot enough for industrial processes. This flame is used to heat homes, an easy requirement for such a hot flame. The heat in the home eventually escapes outdoors where in a large city an urban heat island is created. The flame is about 1500° Celsius hotter than the surrounding environment. The home is about 30° Celsius warmer than the outdoors on a cold day. The urban heat island may be 2° Celsius warmer than the countryside. From 1500° to 2° is a large drop in concentration of heat but since energy can be neither created or destroyed, the lower temperature difference is matched by the increased dispersion of heat to a whole urban area. These are the workings of the second law in action.

In many ways, industrialization has applied the principles of the second law of thermodynamics to *everything*. From the mines to the junkyards, the materials become less concentrated, more jumbled up with other materials, and spread far and wide over the face of the planet. If energy cannot be destroyed, no material can be destroyed. When an industrial product is thrown away, it only disappears from a common line of sight because in truth, there is no away.

LINEAR INDUSTRIALIZATION CREATES PROBLEMS AT BOTH ENDS OF THE INDUSTRIAL PROCESS. HIGH-QUALITY RESOURCES ARE BEING DEPLETED AT ONE END OF THE PROCESS AND WASTE PRODUCTS ARE PILING UP AT THE OTHER. EVERY INDUSTRIAL PROBLEM OF IMPORTANCE IS A PROBLEM EITHER OF RESOURCES OR WASTE.

What is worse, the economic systems in industrial countries put a clock on the time it takes for a resource to become waste. Gross National Product (G.N.P.) is a measure of how *fast* natural resources become waste. At the very time when industrialization is confronting the problem of resource limitations and waste disposal, the economists are proposing solutions that only accelerate this process. This will only serve to make the problems worse.

The solution is obvious. The waste outflow must be converted into a resource asset. Sometimes this process is called recycling—a term that should be deliberately avoided because it conjures up pictures of Boy Scouts on newspaper drives. Newspaper drives may be a perfectly fine thing for children who must learn the concepts of waste management, but recycling on that scale is not a win situation from an environmental standpoint when the burning of fuel in the scoutleader's station wagon is factored in. The ugly truth is that after fifteen years of recycling talk in the United States, the most effective recycling mechanism is the garage sale. Converting the wastes of industrialization into industrial assets is a problem far beyond the grasp of the Boy Scouts or garage sales.

Closing the industrial loop is a project of similar magnitude to the industrial revolution. Undoing the damage is an even bigger problem than industrialization. About one-third of the world's population has been beavering away at the creation of the industrial infrastructure of the planet for the last 150 years with occasional setbacks from warfare. A problem larger than that should mean one thing: unemployment will cease if the resource-to-waste loop is closed because there is plenty of work that needs to be done—so much work, in fact, that it boggles the imagination.

Take the problem of toxic waste. Americans are currently producing such waste at the rate of 300 pounds per person per year. The military, medicine, and the chemical industry are primary villains in this tragedy but no one is innocent. Small businesses such as dry cleaners, print shops, refinishing shops, repair garages, and photo studios are guilty of putting solvents and other hazardous chemicals in to the water system. The computer industry, supposedly a clean industry because it does not have the belching smokestacks of the steel industry, uses chemicals that are so dangerous that belching smokestacks suddenly look like a minor irritant.

Some protest that they are not a part of the problem, but even on an individual scale, it is almost impossible to be a part of an industrial society without being a part of the toxic waste problem. A homeowner paints some woodwork, cleans the brush with solvent, and rinses the solvents down the sewer with the leftover paint. A car owner removes the bugs and tar from the vehicle with a petroleum-based cleaner and hoses it down the gutter. The average homeowner has dozens of materials in the house that have no known disposal method that does not add to the problem of toxic waste. Most waste, unfortunately, ends up in the global water supply: groundwater, creeks, rivers, bays, and the oceans.

These are examples of direct contributions to the toxic waste problem. Many problems are indirectly caused. Buy a car and the paint job, the metal pickling, the rustproofing, and a host of other toxic waste problems have been incurred in the carbuyer's name. A person who uses a bicycle for transportation has caused less environmental damage than a car owner, but there is still damage.

Bicycle manufacture involves smelting, solvent use, rubber processing, painting, and plastic fabrication—all processes that cause toxic waste. It is literally impossible to avoid being part of the toxic waste problem and live in an industrial state.

More importantly, toxic waste problems are out of proportion to the sheer amount of waste being generated. One gallon of gasoline spilled into the water supply will render 750,000 gallons of water unfit for human consumption. The multiplier effects of toxic waste are scary. Remember, it is the water supply, one of the most basic requirements of life, that is threatened by such waste.

Proposed solutions to the toxic waste problem have been, to put it mildly, inadequate. Humans have only begun to understand the magnitude of the problem. Talk of cleaning up toxic waste dumps means digging up the barrels of waste and putting them somewhere else. The fallacy of such a solution is that not all the waste is recovered from the original site and the second site is typically no more ready to contain the waste than the first. Such cleanup efforts mean that instead of one hazardous waste site, we now have two. The real solution to the toxic waste problem has two parts. First, undo the damage already done; and, two, stop the practices that create the problem.

Undoing the damage that has already occurred is a huge subject—clearly beyond the scope of this discussion. Moreover, it is futile to discuss fixes for problems of the past if the industrial state is going to proceed at an even greater pace to create more problems for the future. If humanity can stop the problem from getting any worse by creating new industrial practices, solutions will grow from this effort that will be applicable to undoing the sins of the past.

AN END TO THE ECONOMICS OF WASTE

"An ounce of prevention is worth a pound of cure." Benjamin Franklin

BENJAMIN FRANKLIN SAID many intelligent things. None is more applicable to the twentieth century and its industrial problems. The real solution for the problems of waste is to refrain from making waste in the first place. Unfortunately, this seems at first to be impossible. Efforts to stop the motion toward a fuller realization of the throwaway society have failed because too many people vote for such a lifestyle in their patterns of consumption. Moreover, the problems caused by paper and bottles and cans are not really that important. Paper is more or less biodegradable, bottles are environmentally much the same as sand, and cans have enough value to be picked up. The real problems are caused by the waste that is harmful in the sense that there is no safe disposal method. These wastes, which will ultimately doom humanity, are the leftovers from products and processes that support human life. Such processes cannot be stopped, they may only be altered and redesigned.

Since it is impossible to stop toxic waste by stopping the industrial activity responsible, the only reasonable solution is to redesign the industrial process so that the waste itself is addressed. There are many ways to do this but each method has one common theme. The industrial process must be redesigned to eliminate whatever waste is generated.

Redesigning the industrial process to eliminate waste has been tried in many industries. Waste usually lowers profits and the economic incentive to eliminate waste is very strong where this is applicable. Basic industries such as plywood manufacture and meat packing are models for this form of efficient behavior. In such cases, the raw materials enter one end, finished goods emerge from the other, and very little is lost in the process. Meat packers claim that the only part of a hog that is not used for commercial ends is the squeal. One plywood maker claims the only thing lost is a little smoke which is mostly activated charcoal and he is looking for a way to market that. Meat packing and plywood manufacture are very mature industries—some of their processes are centuries old. It stands to reason that by now most waste in the process has been eliminated and almost no toxic refuse is produced. Younger industries have no such advantage. Large industries, such as the chemical industry, which have the political clout, avoid the toxic waste issue by getting a license to pollute. Individuals or small businesses, such as the dry cleaners, quietly dump their waste improperly and hope that their small size keeps the concerned authorities from noticing them. Both have taken the action because the costs of proper waste disposal are beyond their means or will.

Ignorance and poverty are reasonable excuses for the behavior of small businesses and individuals. No such excuse exists for large-scale operations. A chemical plant has many chemists on the payroll who have all the expertise necessary to determine whether a form of waste is hazardous and what can be done about waste. The problem with the chemical industry is that many of their commercial products are toxic substances with a high probability for improper use. Even if a substance such as acrylic enamel catalyst were to come with detailed instructions for environmentally benign use, most auto body shops are not equipped to follow such instructions. This example of catalyzed acrylic enamel is apt. Flawless paint jobs are more than an aesthetic pleasure. Paint makes a car last longer.

Since the automobile will be a necessary part of the transportation system for the near future in North America, and it is virtually impossible to make a car without making an environmental mess, then it follows that making fewer cars that last a longer time is in the interest of biosphere. Getting the paint job right at the factory is a relatively simple task, although the painting equipment is very sophisticated. Lasers are used to assure uniform paint thickness throughout the entire coat. Fixing a slightly bent car is harder. The paint on the car may have faded and will be hard to match. The work must be done in dirtier, more primitive surroundings. Although the paint in the factory can be baked on under very high heat which causes the paint to level, such heat cannot be used in repair work. The assembled car has plastic parts, both inside and out, which will melt if heated.

Not to worry, the chemical industry has responded with a host of very sophisticated products to make possible the job of body repair. Acrylic enamel is the paint of choice for this work. It has the advantage that small errors, such as runs or sags, can be corrected before, or even after, the customer notices them. Enamel is only glossy on the surface that dried in the air. Carefully rubbed-out mistakes can be waxed to look shiny but they have microscopic cracks that can become rust spots over time. Since avoiding rust is the real reason to paint the car, this is unacceptable.

Cracking is solved by using a catalyst that makes the paint cure chemically instead of losing solvents into the air. This means a shine can be had at rubbed-out levels beneath the original surface.. Once the catalyst is mixed into the paint, the process begins. Paint not used becomes like hard rubber in the bottom of the can, even if the can is sealed properly and the paint is three inches thick—great, if somewhat spooky stuff. The catalyst is so toxic that inhaling a few breathfuls can make a person very sick. The can is clearly labeled "FOR USE BY PROFESSIONAL PAINTERS ONLY" on the tenuous assumption that everyone in the body shop has been briefed on how to handle catalyzed paint. If the auto painter has been trained in a vocational training school and was not sleeping the day it was brought up, he might know that he must have more than a particle mask. He must employ a complete face mask with outside air, but the nature of his work means that even equipped with a proper mask, the body painter is about to become an ecothug and has no choice in the matter.

The painter first turns the paint into a very fine mist and sprays it into the air. The better his spraygun, the finer the mist. The target, of course, is the car but about 40 to 50 percent of the spray misses, bounces off the surface, or otherwise enters the atmosphere. The spray is caught by the filters in the spray booth but many of the volatiles go right through the filter and out into the air where they are eventually washed into the water supply by the rain. What does the body shop do with the filters when they are full? They can be burned which puts toxic chemicals into the air and back into the water. Or they can be buried where eventually the chemicals trapped in the filters will leech into the groundwater. The body shop is trapped. It is doing the public a service, which extends the useful lives of their automobile, but the body shop is an environmental disaster happening on a daily basis.

There are only two routes out of this dilemma. First, the chemical industry could reformat the catalyst and other materials so that they are both functional and environmentally benign. This is the preferred strategy since it is easier to cope with the problem in a central location such as the chemical works than to try to enforce sound environmental practices in hundreds of thousands of body shops. Second, the chemical industry could come up with a second chemical that would neutralize the effects of the first. This is sometimes the only solution that works and environmentally aware companies have begun such practices. Chemicals for printing pictures include acids and other poisonous substances. The environmentally conscious photographer can purchase chemicals with neutralizing agents that are added to solutions before they are washed into the sewers and into the water supply. What percentage of people buys these chemicals and what percentage carefully neutralizes solutions is anyone's guess, but it is a solution that is at least possible.

These two approaches illustrate the difference between UPSTREAM and DOWNSTREAM design. Upstream solutions are always preferable because problems are solved before they are created. Downstream solutions, by which waste products are treated before disposal, are available but rely too heavily on humans becoming sinlessly perfect. Unless this is a possibility, downstream design solutions should be avoided, but both approaches represent improvements over current practices.

Permutations of the notion of upstream design solutions are the first real light to shine on this litany of disaster. Upstream solutions include the notions of permanence, recovery, and fit.

UPSTREAM DESIGN is manifest in a desire to build permanently. There is an economy of the well-designed and well-built. Beautiful buildings and other forms of permanent structures built by humans are endowed with a form of life. They can almost take care of themselves because they attract people who wish to see them preserved. From Notre Dame in Paris to the Brooklyn Bridge, great structures are almost immortal because mortals will rise up to preserve, rebuild, and protect them. Tearing down old structures and building new ones consumes more resources and creates more waste than maintaining a structure through the centuries. In the rush to settle North America, this lesson was left behind in Europe. Permanence is the most basic form of upstream design in that building well the first time is the simplest way to save resources and eliminate waste. Thinking that something is to be built as a permanent fixture has a profound effect on how a building is built. Permanence is industrial immortality. The thought is sufficient to alter every effort from the first line on the paper to the last swing of the hammer.

Permanence is a function of design. Design as beauty is essential to permanence. Committees are rarely formed to save eyesores. Design as function can assist in the goal of permanence by making maintenance easier. Design as beauty is the easiest to explain because everyone can think of an example. Builders for centuries have known this trick. Design it to be beautiful and the powerful will insure access to premium materials and labor. Build with good design, premium materials, and skilled labor, and the result is the Acropolis in Athens, the Vatican, or the Eiffel Tower. Apply this principle to a city and you get Paris, Prague, or Copenhagen. The only reason to restate such an obvious notion is that one cure for industrial waste must be permanent structures. No law has been found that says a powerplant, or sewage treatment plant, or waste recovery center must be ugly. Humanity must arrive at the conclusion that this is the only planet for light-years in any direction that will support human life. Structures necessary for human survival will be with us for a very long time and we are going to have to look at them. They may as well be beautiful.

Of course, not everything can be made for permanence. Permanence is an elusive goal and not everything made can be made for extended life. Everything is subject to deterioration. Quality materials can slow this process but cannot stop it. The real secret to permanence is maintenance. Europeans understand this but North Americans have trouble grasping the concept. Maintenance is considered low class work, though the skills necessary to rebuild a structure are more varied and difficult to master than those necessary to build new structures. Most cities, roads, bridges, water systems, and other industrial installations in America are young. As these permanent installations began to show their age in the 1970s and 1980s, Americans responded with surprise and annoyance. Worst of all, money was cut for planned, routine maintenance. At the very time when a maintenance ethic should have begun to emerge in the North American industrial cultures as their societies matured, maintenance became a dirty word.

Maintenance is not culturally entwined with American culture. Quality, if it is referred to at all, is taught in the context of the poem called "The Deacon's One-horse Shay." In the poem, there is a vivid description of all the pains taken to make the perfect shay. Every part was fussed over. The resulting shay was quite wonderful and it lasted 100 years when all the parts failed simultaneously. It would probably be better to inform children that George Washington's ax may be on the second head and the third handle, but it is still George Washington's ax. The shay example demonstrates no maintenance: the ax example glorifies it.

Modern examples of this cultural phenomenon are easy to see. From car batteries to power tools, the notion of a maintenance-free product is routinely cited as a form of quality. With batteries this may be true, with power tools the issues are not so clear. European tool makers design their tools to be easily disassembled for repair and maintenance—a fact that troubled the North American buyer when such tools were first marketed. Europeans were forced to re-design their tools to be "maintenance-free" for the North American market. The American buyer was questioning the quality of a product designed to be repaired. (Keep in mind that maintenance-free, as understood in this context, means that the whole product is disposable for a smaller failure.)

The truth is that a product which can be repaired easily has a much greater chance of permanence because only the part that fails must be replaced. The quality that comes from premium materials and careful construction is no match for a similar product to which the quality of maintenance and ease of repair are added. The failure to glorify maintenance may be the most difficult cultural barrier to permanence in North America.

Upstream design can manifest itself in industrial goods that make no pretense toward permanence when they are DESIGNED FOR RECOVERY. If bridges, water systems, and highways are the industrial products most likely to be designed for permanence, packaging is the industrial product least likely to be so designed.

According to current design definitions, packaging is intended to protect something more valuable than the package itself. Packaging is supposed to be as valueless as possible and still do the job of protection. When the job of protection is finished, the packaging has lost its design value and is valueless and disposable. It should come as no surprise that packaging is a major contributor to the waste problem.

The packaging industry is more sophisticated than the definition of protection would indicate. Because packaging has become a form of marketing, the notion that packaging is protection is very often overlooked. Colorful, eyecatching bubblepacks must still be less valuable than what is under the bubble or the economics of packaging fails to apply.

There are some rare exceptions to this principle. Parents will buy an expensive toy that the child found totally uninteresting compared to the box it came in. Rest assured, however, that from the manufacturer's point of view, the box cost less to produce than the contents. Even so, the point is made that although the package may have less value than the contents, it is not totally valueless. Many forms of packaging are pressed into further service after their initial design use has ended. Homeowners have found thousands of uses for empty coffee cans. I know from childhood experience that a generator packing crate makes a terrific treehouse.

Modern industrial existence owes much to the packaging industry. Medicine owes much of its success to the possibilities of sterile packaging. Fragile electronic goods from all over the world can be bought almost everywhere because of packaging. Most importantly, the food distribution system with its freezing, bottling, and canning has allowed for huge cities to exist far from the sources of food. Modern life without modern packaging would be primitive.

Packaging is not an innocent actor in the environment. It is the source of much of the solid waste generated by industrial societies. It is an environmental problem of the first order. The solution to this waste problem is to remember that packaging is like many other forms of waste in that it has value. Recovering that value should help pay for the recovery efforts in an industrial and economic sense. Already, aluminum cans have become worth recovering because of the cost of making aluminum from bauxite. Smelting aluminum takes a great deal of electricity and rising energy costs have made recycled aluminum cheaper than the aluminum from ore. As the costs of throwing things away become known, it is economically clear that is not only high-value metals such as aluminum that can be recovered, but that everything currently considered waste must be recovered.

The industrial loop will be closed when humans figure out how to recover all waste products. In this respect, industrialization must mimic the ecosystem. The study of the environment is the study of one system which feeds off the other. That the planet is a closed ecosystem is both fascinating and frightening. That the millions of plants and animals can be organized into a loop where one species' waste is another's food is more than an interesting phenomena, it is an industrial mandate. Until this closed loop organization found in nature is copied by industrial design, the planet is doomed. As both are necessary for human survival and both are internally interdependent, both must be similarly maintained. A planet organized so that every waste product must be accounted for in the environmental books cannot tolerate one species that believes it can continue to operate in a linear fashion where the process is glorified and plunder and waste are counted as externalities. Human industrialization is only half complete and doomed as a system until it copies the natural processes.

What makes recovering all waste an upstream design imperative is the notion of fit. The difference between a skilled billiard player and the first-time player is the ability to control the cue ball. The novice player is so concerned about making the shot that the path of the cue ball after the shot seems unimportant. The skilled player has learned that when the cue ball stops in the right place, making the next shot is easy. Modern industrialization can be likened to the first time player. Because it is so engrossed with manufacturing and marketing the product (making the shot), industry rarely concerns itself with environmental problems (worrying about the path of the cue ball). The concept of fit can be understood in the context of the leave in billiards. The difference between industrialization built on a closed loop system that mimics natural systems and the current form of industrialization is as great as the difference between the two billiards players.

Waste can be recovered in many ways. Waste can be recycled as in the case of aluminum cans. Waste can be burned for its heat content. Then there are the imaginative ways. The digestive systems in cows can process cellulose. Hay has cellulose which makes milk and eventually hamburgers as it is consumed by the cow. Interestingly, so does cardboard. Ground packaging could be fed to cows as food, except for one problem. The ink used in printing labels, and the residual chemicals from the process of making the cardboard, render milk and meat from cattle fed such waste unacceptable for human consumption. If nontoxic packaging could be printed with chemicals and inks that could be safely processed by cattle into meat and food, the day could come where humans will drink the milk and feed the carton back to the cow, who in turn would make more milk and fertilize the earth. The decision to use a vegetable-based or some benign ink in the packaging would be the key to making such a closed loop system possible. If packaging were designed so that it could become food for animals, it would be a perfect demonstration of the concept of fit as the waste of one product neatly dovetails into another.

Closing the industrial loop will be a very complicated and difficult undertaking. The path from wasteful and plundering linear industrialization to a system where all raw materials and waste products are accounted for will be long. Because the current system of industrial organization is a failure on the verge of collapse, a solution is necessary and a closed-loop industrialization scheme will solve the problems. Moreover, closed-loop industrialization will require generally the same organization and skills of linear industrialization. A solution that requires a violent military revolution is not a solution. Closed-loop industrialization, or industrial environmentalism is a solution which requires the smallest change in human behavior. The problems of industrialization are so severe that there is not time to fool around with revolutions. The essence of industrial environmentalism is total resource recovery. When total recovery becomes the goal, permanence and fit are the guiding strategies. The enormous size and scope of such a conversion to industrial environmentalism should not be viewed as a liability, but the prime asset. Building the way out of the absurd and damaging legacy of linear industrialization will provide employment for several generations.

CHAPTER ELEVEN
TECHNOLOGICAL ELEGANCE

Those who believe that there is a difference between art and technology know little about either! Anon.

THE BATTLE BETWEEN industrialization and environmentalism is more intense than the conflict between communism and capitalism and 6000 years older. It is as old as the conflict between the hunter and the farmer. By asserting that the only route out of the industrial-economic crisis is to join these two opposing forces, this book could easily be dismissed as a utopian fantasy. Industrial environmentalism is an idea whose time could never come! Moreover, because cooperation between the industrial and predator classes has produced such lovely phenomena as Fascism and the military-industrial complex, the suggestion of cooperation is frightening and fraught with danger.

Yet, industrial-environmentalism is neither a utopian fantasy nor an impending nightmare. It is an exercise in social, economic, and political minimalism—a solution that requires the least change in human behavior. The economic-industrial crisis is the creation of human efforts. The efforts that caused the problems can and must be redirected toward a cure. Industrial environmentalism is not a utopian fantasy because both industrial activities and environmental activities can be undertaken jointly as a means for survival.

A common motivation for survival means that these two strains, often considered dissimilar, are really alike. There is a much misunderstanding between those who consider themselves industrialists and those who would be environmentalists, to be sure, but this lack of understanding between these groups can be minimized if both parties understand that they are pulling in the same direction. The misunderstanding could be eliminated if this move toward a common goal could include a common strategy. Industrial environmentalism is that common strategy.

The best way to understand industrial environmentalism is to think in terms of a total environmental accounting. Everything that is made must either be made for permanence and maintenance or must be designed to be unmade at some future date. This goes far beyond what even the most radical environmentalists have ever proposed. There is a big difference, however, since most environmental proposals so far have been negative: stop cutting down trees, strip-mining the valleys, polluting the air and water, and selling parkland to developers. When fuel was short, the nation was asked to slow down and turn down thermostats and shiver in the cold. Stopping the damage to the environment was going to cause pain, it was thought, though if the pain were felt by closed mining communities and steeltowns, truck drivers, old people and welfare families, so much the better. Environmentalists have never quite shed their elitist image because of their negative suggestions affect groups less powerful than they.

Industrial environmentalism is positive and inclusive. Because the industrial crisis is caused by errors in design and construction of the industrial infrastructure, rebuilding should be considered an economic development challenge for the industrial nations. It is a job large enough to occupy the energies of a generation or two. Building is more positive than stopping and has the advantage of including groups that are excluded from most environmental proposals. Building a closed-loop industrial system to the most stringent environmental standards will wed industrialization with environmentalism and is a positive, inclusive economic strategy for survival. It is also a method for making nations very rich.

ELEGANT TECHNOLOGY

BECAUSE HIGH TECHNOLOGY has become a meaningless phrase, elegant technology is the best expression for the technology necessary for industrial environmentalism. Elegant is a description used by designers and engineers to describe a design that solves more than one problem. If a new design is cheaper to manufacture, is more durable, and opens a distribution bottleneck, the design is more elegant than the design it replaces. Other design criteria for elegance are: reduces waste in manufacture, uses a plentiful natural resource, comes from a more reliable source, is the waste product of another process, reduces the energy necessary for manufacture, reduces the tooling costs, and is more beautiful. Those who would claim that beauty is not a common element of engineering should look at the precision castings inside engines such as those on the Italian motorcycle, the Moto Guzzi. The Italians have been at the casting business since before Michelangelo and they have learned a great deal in 500 years.

Industrial environmentalism asks that elegant engineering add relevant criteria to the accepted definitions. To be elegant, a design should include a provision for its ultimate disposal. A new design must deal with the waste generated by the manufacturing process and the product itself. The person who designs a product is most likely to have the relevant information as to how it can be safely disposed of.

Elegant technology is that technology where the disposal, and any other form of downstream problem, is dealt with from the beginning. When industrial designers acknowledge that raw materials may only be borrowed from the planet and must be returned at some future date, then and only then will technology be truly elegant. The most simple and direct criterion for evaluating elegant technology is to ask, "what will happen when *everyone* uses this technology?"

THE RULES OF INDUSTRIAL ELEGANCE

THE FOLLOWING EXAMPLES of elegant strategies for industrial environmental renewal are just that—examples. A technologically literate reader can think of dozens of other examples and criticize the choices made.

Examples are debatable—but the essential principles of elegant technology are not. Like Murphy's Laws, there are corollaries yet to discover. Even so, while the following rules may not be definitive, they describe most industrial-environmental applications.

EVERYONE IN AN INDUSTRIAL SOCIETY IS A PRODUCER.

This is difficult for most people to understand because what most folks produce is called garbage. There is a Japanese production principle called Just-In-Time (JIT) production that helps illuminate this rule. JIT works best when every worker in an industrialized setting acts as if the next worker or process is a customer.

The consumer of industrial products would benefit from such thinking. Everything that is consumed is being processed for the next customer. Although it is difficult to imagine you as a consumer are producing a product for the trash hauler, that is really the correct way to view the relationship. Just as in JIT, where the quality of the ultimate product depends on each person doing the job correctly, the harmfulness of waste is a function of each person's processing efforts. Of course, because of the complex nature of most industrial products, the average consumer is totally lacks the means of processing products into useful forms of garbage—even with good intentions. A consumer's inability to process goods into waste correctly is determined by production decisions made much closer to the original raw material.

POLLUTION IS A FUNCTION OF DESIGN.

The most effective way to treat pollution burdens is not to create them in the first place. Source reduction in the industrial setting may be, in fact, the *only* way of eliminating industrial pollution.

Design is hard to understand in an economic setting—especially for an American brought up on the notion of designer jeans. Designer jeans are to design what a comic book is to the Bill of Rights. Even a person who understands the economics of why designer jeans command a higher price knows that all jeans are equally difficult to dispose of properly when worn out. Design as a solution for industrial waste seems hopelessly subtle. It is not. Simply put, if all waste is to be properly processed, the separation of waste into components is almost as difficult as the original combining of elements. Demanufacture is roughly equal to manufacture.

Demanufacture is made easier when it is anticipated in the original manufacturing process. Designing for Disassembly (DFD) is best done by the original product designers. They have the most relevant expertise.

IT IS ONLY MONEY!!!

The value of money, like pollution, is a function of design. Design is the key *link* between the economy and the environment. The financial superstructure must be the servant of the real economy—not the other way around! Despoiling the environment so some predatory financiers can prove their industrial ignorance while misprogramming computer chips is intolerable. The argument that something of environmental necessity cannot be done because there is not enough money, is utterly absurd. Money can be created by pushing a few buttons on a computer. To think otherwise is to be trapped in a preindustrial mind-set. If monetary policy creates pollution problems, then the time has come for a new monetary policy.

The principle of elegant designs can, and should, be applied to everything. Since everything is too wide a subject to treat meaningfully, this discussion of the applications for elegant technology will be confined to five areas: waste management, food production, medicine and population control, elegant tools, and energy. Keep in mind that because elegant technology is closed-loop technology, these five areas cannot be easily separated, even for this simple discussion.

ELEGANT WASTE MANAGEMENT

THE MOST IMPORTANT ASPECT of waste management is to separate toxic substances and combinations from the rest of the waste stream. Cardboard can be fed to four-chambered-stomach farm animals as food if it is not contaminated with toxic inks and other chemicals. Burning waste products is less possible when toxic waste is present. If toxic substances can be kept from the rest of the waste products, many interesting possibilities for elegant solutions for waste disposal become possible.

Nontoxic wastes can be placed into three categories: fresh organic waste such as human or animal excrement, grass clippings or leaves, and discarded food; fibrous or flammable organic waste such as wood products, paper and packaging, or tires; metals and other inorganic waste such as old concrete. Most non-toxic wastes have intrinsic value so that recovery efforts should pay for themselves by almost any accounting method, even (or especially) if the recovery methods are sophisticated. Some waste disposal methods are profitable.

Fresh organic waste should be composted and returned to the soil. The Dutch have applied significant technology to mass composting efforts. They believe that composting creates a fine fertilizer that will replace a much petroleum-based, synthetic fertilizer. Composting is equivalent in hard costs to any other form of sewage treatment. Living in a tiny country, much of which was wrested from the seas with great effort, the Dutch long ago reached the conclusion that fresh organic waste is too valuable to discard—even if they had a place for disposal. They realize that proper composting is best done by professionals, with the proper equipment, and in a central location. Farmers may compost for themselves because they have the space for such an activity, but crowded cities afford no such option. Even in the United States, where many people live on suburban plots large enough for personal composting efforts, such effort is rare because it is a bother. Waste to fertilizer conversion will only happen in urbanized, industrial societies when it is done in cities and towns because of the nature of the people who live in them.

Burning waste, even when the heat is fully used, is a solution of last resort. Sophisticated burning efforts can produce safe gasses such as carbon dioxide and water vapor—both used by plants. The ash has significantly less mass than the products to be burned and can be mixed into the compost heap. Ash disposal is greatly simplified if the burning process does not itself cause the formation or collection of toxic substances in the ash. Therefore, a burning solution must involve upstream considerations.

The burning process itself is a downstream solution that is greatly simplified if what is to be burned lacks known toxic additives. Plain paper can be safely burned in the environment. Paper with ink on it, which is most of the paper discarded, presents a much greater burden. Plastics are flammable but usually produce toxic gasses when burned. The best solution would be to make ink and plastics so that they may be safely burned. When that is not possible, the incinerator must be more sophisticated to compensate for these substances. As in all good design problems, this decision will involve a trade-off between the costs of reformulating ink and plastics versus the costs of greater incinerator sophistication.

Since the public will end up bearing the costs for any solution, including the decision to do nothing, how this is accomplished and by what method must be a public decision. Since there is valuable heat as a by-product of this disposal method, the benefits must also be weighed in the cost accounting.

There is no known safe method for disposing of automobile tires. There is much potential energy in a worn-out tire but burning them causes a big mess. Whoever solves this one will be an environmental hero.

One thing must be remembered whenever burning is suggested as a solution. Burning is, by definition, rapid oxidation. Simply, this means that while visually the pile of ash is smaller than the pile of garbage, oxygen has been added to whatever has disappeared. Eventually, most waste plus oxygen equals carbon dioxide. Of all the pollutants, carbon dioxide is the least harmful, unless huge amounts are generated. Any burning scheme must consider the *volume* of carbon dioxide generated and plan for its reuse. Societies must be very careful because fire, by itself, will not eliminate any environmental stress.

Metals should be recovered and recycled into the resource stream. Old concrete and other discarded building products can be safely buried. Buildings are *supposed* to be permanent so their disposal does not cry out for an elegant solution. Steel can be taken out of the waste magnetically and most other metals have enough value to justify individual recovery efforts.

The key is separation. First there is the separation between toxic and nontoxic wastes and then there is the separation within categories. Who will do the separation is a social question of the first order. There are several answers. In a community where people are normally very responsible, trustworthy, thrifty, and fastidious, the separation efforts may be entrusted to the individual waste generators. An apartment building could have two garbage chutes: one for flammable waste and the other for metals. Fresh organic waste would go down the sewage system. Such as solution would be the least expensive but the presence of a very few uncooperative individuals would cause the scheme to fail. A simple-technology alternative would be to hire unskilled workers and separate the waste at a central location. The snazzy method would be to separate the waste using highly automated systems.

A technologically simple solution would be best for locales with unemployment and lower capital availability. The automated method would be best for wealthy suburbanites who like to see their city employees wearing ties. How the separation process is accomplished is far less important than the critical decision to separate.

Coping with nontoxic waste is best solved with macro solutions. The toxic waste problem is a micro problem in that each chemical or other waste product must be treated individually. The problem of nuclear waste is being deliberately ignored here because there is no known solution. The great geniuses who gave humanity this form of toxic waste must be put to work on a solution—if there is one.

It must be remembered that toxic waste is both a technical and a social problem. The technical problem can only be addressed with the certain knowledge that everything that can be made can be unmade. Some toxic waste can be burned under very careful conditions if the true nature of the waste is known. Even this solution assumes a great deal about the social reality. For toxic waste to be known, collected, and treated properly, it must be tracked very carefully from manufacture to disposal. Getting a large company to spend the effort to track the progress of a toxic waste will be very simple compared to tracking the millions of individuals and small companies who produce toxic waste. It is impossible to imagine an environmental police force able to check every person that buys a can of paint remover. It may come to the point, however, that toxic substances are sold only to licensed individuals and companies paying to neutralize, alter, treat, and recover the toxic waste. This would force companies that make toxic substances to alter their formulas or get out of the market—a powerful incentive to reformulate their products. This would be the preferred upstream solution.

Unmaking a toxic waste involves a process *very roughly* equivalent to making the product in the first place. Under linear industrialization, the costs of toxic de-manufacture are nearly impossible to ascertain because much is still technologically impossible—under such conditions, costs are practically infinite.

Early DFD efforts have shown that with elegant design, demanufacturing requires approximately 1/4 the energy and 1/10 of the labor necessary to manufacture the original product. To be completely successful, DFD should meet a disposal cost target of 1/100 of the energy and 1/1000 of the labor.

For many vital toxic substances, DFD may prove impossible. Unless a toxic waste can be converted to a valuable new resource—a very improbable solution in many cases, or can be easily identified and neutralized, proper disposal for toxic waste will, mostly, result in added costs. These costs, whatever they may be, must be reflected in the selling price of toxic substances—however necessary.

ELEGANT AGRICULTURE

AGRICULTURE IS THE ORIGINAL INDUSTRY from which all other industries sprang. Not only are many tools, inventions, systems, and production structures directly related to agriculture, but agriculture provides the food necessary for the growth of all other industries. In a real sense, cities are the product of agriculture. Industry and agriculture are so closely linked that a discussion of one is a discussion of the other.

If environmentalism is defined as preservation, then agriculture is the original anti-environmental activity. Agriculture does not preserve but alters the landscape. Whether this alteration is an improvement is very much open to debate. Some agricultural practices, such as terracing, are improvements over the natural state. Others, such as slash and burn—unless done in a carefully controlled, preindustrial, long-cycle manner—leads to the environmental catastrophe of the Amazon basin.

There is an incredible array of agricultural practices. Almost every method of food production has probably been tried somewhere on the planet. Agriculture methods vary with soil conditions, climate, and culture. The *environmental* reason for owner-operated agriculture is that because these natural differences are so subtle, each small plot of land must be cared for like a cherished member of the family. Within the body of collective wisdom gathered through experimentation are to be found all the necessary elements for an environmentally benign, elegant, and sustainable agriculture.

Environmental decline caused by agriculture can be traced to two basic sources: the adaptation by agriculture of the lessons of linear technology learned in the industrial revolution; and, the scale of agriculture brought on by the rapid population growth rates associated with the twentieth century. Both sources are closely related and manifest themselves extensively in the agricultural practices of North America. Because North American agriculture is considered to be so successful, it is widely copied and admired throughout the world. Without meaning even the smallest slight to agricultural practices in the rest of the world, this discussion will focus on industrialized North American agriculture. The environmental questions of North American agriculture must be solved or else we are leading others to their doom.

As the early products of twentieth century industrialization found thier way to the farm, life became much easier. Tractors and the internal combustion engine joined forces with nineteenth century inventions such as the Deere plow and the McCormick reaper to make farmers faster and more productive. Farms that could be run by a single family grew in size. The cultural definition of a farmer also began to change.

Each additional industrial product changed the way farm work was organized. If a farmer bought a tractor, his job description changed to include the ability to fix a tractor when it breaks at a critical time during harvest. There are implement repairmen in the small communities that serve agriculture, but when harvest comes, they are too busy to help everyone. Farming as an occupation now includes the skill of heavy equipment repair.

The distinction between modern and preindustrial farming is the availability of electricity. A modern dairy barn is filled with the technologies of a small factory—compressed air, hydraulics, refrigeration, vacuum lines, and conveyor belts. Equipment for bovine medicine and bacteria reduction, artificial insemination, and equipment sterilization show that the modern farmer has become a scientist. To feed the cows, the farmer must grow crops—a process that involves selecting genetically altered seeds, testing soils, purchasing appropriate soil nutrients and predicting the weather.

The modern American farmer-mechanic-lab technician-scientist-businessperson must have absorbed more information and mastered more skills than almost any other known occupation. Far from being peasants, modern farmers constitute an elite profession. Less than two percent of the population feeds the country with surpluses for export—the agricultural crisis of the 1980s was *not* caused by inefficient production methods. Not all this production efficiency flows from good farming practices, unfortunately. As with other industrial enterprise, American farming is charged with environmental devastation—such as soil erosion, groundwater contamination, and pesticide poisoning.

If an elegant agriculture is the goal, it must be remembered that the cultural needs of agriculture are larger than the technological and environmental changes required. An elite profession is being asked to get *better*. The science of sustainable agriculture is well researched, yet the 1980s crisis in American agriculture has increased the cultural difficulty of converting to sustainable agriculture. During that time, anyone with debt faced the fact that banks could change numbers faster than a farmer could increase production. The old song about the mortgage working overtime gained new meaning. When the numbers exceeded production, the lender foreclosed. The farmer most likely to be in debt was young. The 1980s catastrophe in American agriculture changed the demographics of the land—there are no young farmers left. Most operating farmers are in their 50s, 60s, and 70s. These old foxes are set in their ways. If this trend continues, we may see the day when potential farmers must be financially encouraged to farm—much like rural doctors. It is possible that the United States will have a sustainable agricultural future only if elegant agriculture is treated as a public works project.

Any form of enterprise that has many examples of linear technology is probably going to be an enterprise that has replaced human muscle with mechanical power. Nothing could be more descriptive of agriculture. North American agriculture is much admired because it is productive. This measure of productivity is not a measure of how much food is produced on an acre of land although North American agriculture rates very high in this category; rather productivity is a measure of how much food is produced by each farmer. By this measure, the North American farmer has no peer. It must be remembered that this form of productivity is a function of automation and automation requires energy. The high energy requirement for the current realization of North American agriculture is a technology trap.

Primitive agricultural methods avoid the technology trap, but primitive agriculture means giving up many things that are routine fixtures of twentieth century life and choosing a life that is very difficult. Some groups, such as the Amish, have maintained a continuous link to past agricultural practices and have an organized social structure that eases the burden through communal cooperation. Most modern farmers have long since passed the point of returning to past agricultural practices. The ability to weld a broken part on a combine and the ability to fashion a harness for a horse are two very different skills. As multitalented as most North American farmers are, their talents are different in scope and function from the talents of an Amish farmer. To adopt the methods of primitive technology, most North American farmers would literally have to start over from the beginning. As farming is a complex, multiskilled occupation that really can only be learned by growing up on a farm, starting over from scratch would give the expression "born again" a new meaning. This is obviously impossible.

Energy is not the only technology trap. Modern agriculture relies on a host of synthetic pesticides, herbicides, and fertilizers. Anyone who seriously believes that pesticides are something that can and should be eliminated, should be required to read the accounts of the grasshopper invasions of the 1870s and 1880s. People starved to death. People watched as a year's work was devoured in 30 minutes. Synthetic herbicides and fertilizers have caused productivity of both land and human effort to rise dramatically. Fragile, but high producing, hybrid seeds need heavy applications of both. Doing without either could cause serious food shortages.

Synthetic land additives are destroying the very land they are supposed to protect. The rich, black topsoils of the central prairies of the United States are not so rich and black any more. In many places, continuous applications of chemicals have destroyed the natural fertility. In places where topsoil used to contain 200 kinds of worms, there are none left. Worms enrich the soil although some, such as cutworm, destroy crops. Killing them all to destroy the one reduces natural fertility. Add to that, agricultural practices which encourage erosion, and there is not very much topsoil left—even the dead kind. Fertile topsoil is as necessary to life as water. Natural fertility can be replaced with artificial fertility and the plants will not know the difference. Destroying natural fertility and replacing it with a fertility that is manufactured from a finite supply of oil and natural gas is a prescription for disaster. Americans should be aware that Iowa is now largely a natural sponge that holds water, chemicals, and plant roots. Because these synthetics are based on expensive petroleum feedstocks, the first sign of an agricultural crises caused by this technology trap will be widespread economic distress. Economic distress among farmers is merely the symptom of the impending environmental distress, which, in turn, could cause a catastrophic collapse in food production. The society that is caught in the technotrap of synthetic land additives will be crippled.

Elegant solutions to the problems of agriculture are socially possible but technically difficult. Replacing synthetic land additives and fossil energy is a huge problem. Often, the technoproblems of agriculture are caused by practices that are perceived to be the ideal or best solution. Most farmers will not argue substance when the subject of environmental technotraps is raised. Their response is more like, "Sure I have problems because energy is too expensive, fertilizers do not raise productivity enough to pay for themselves, and the poison I am using to kill weeds is making me sick. I'd drop every last one of them if there was a better solution!" Farmers probably understand the notion of a closed loop system better than anyone. Not only are they daily witnesses to natural processes that city dwellers may see a few times in a lifetime, they already employ many elegant, upstream, closed-loop systems. Every time a farmer spreads manure on the field, an age old and simple demonstration of a closed system practice is taking place.

The farmer does not have the time, energy, or resources to organize new industrial solutions—running a 600 acre farm is work enough. It is organizationally impossible for a single farmer to change the base of the fertilizer from natural gas to municipal waste. Composting produces a low-grade fertilizer in the sense that there is not as much enrichment value per pound as in the concentrated synthetics. Shipping the organic waste from the city back to the farm may seem a good idea, and inevitably, that is what must be done.

The costs of shipping heavy, low-grade fertilizers would be absolutely frightening to an industry that has been plagued by shipping woes. The farmer cannot do the research and development necessary to concentrate compost so that shipping would not be a greater hassle than the current levels. That is the job for universities and cities and industry. The farmers cannot pay the price to ship low-grade material. The elegant technical solutions for agriculture must come from cities populated with people who have never thought for one second about the source of their food. City dwellers have the capacity to solve the technical problems of agriculture but first they must become aware that agriculture dilemmas are their own as certainly as zoning ordinances and police protection.

Elegant solutions to agriculture must have an emphasis on a trouble-free operation. Farmers with feedlots could probably generate their own methane with devices such as are used on large farms in China. The Chinese claim that some of their collective farms are energy self-sufficient because of their methane generators. As a typical North American farmer has about the same resource input as a Chinese collective, the scale of the technology is about right. For Chinese methane generators to work in North America, they must be almost automatic and extremely reliable. This cannot be a technology that requires several operators working full time. A farmer could be encouraged to solarize the buildings, make fuel from droppings, or any of a host of other elegant solutions, if and only if the farmer can be assured that the solution is cost-effective and works as promised. Until now, most of these solutions have been nothing but a major headache. Equipment maintenance already occupies more time than the farmer would wish-more headaches he does not need.

Of all the needs for elegant technology, the needs of agriculture are most pressing. There is simply no way to argue with the biosphere and raise food simultaneously. Farming in North America is an industrial enterprise, but it is also an environmental enterprise. Only an industrial environmental solution will produce the required result. Linear industrialization and agriculture are incompatible. Closing the loop with agriculture is not so much a problem of farmers but of joining the farm to the city. An elegant agriculture must move away from a focus on production and toward diversification and sustainability. Government agricultural programs must trade production controls and price subsidies for environmental concerns. This known route of city-farm cooperation should be explored further in any democracy where the numbers of directly interested farmers have fallen so far that all legislative muscle is lost.

Elegant Medicine and Population Control

UNREGULATED INDUSTRIAL MEDICINE in the United States has redefined the meaning of absurd. While millions are denied basic care, others have so much medicine forced upon them they must go to court to have treatment stopped. While prenatal care is neglected, hospitals will spend hundreds of thousands of dollars to "save" the life of a one-pound prematurely born infant—a life permanently deformed by invasive medicine. Transplanted organs promise new life but in fact, make patients permanently addicted to hospitals while incurring costs that no one person could ever afford. Medicine consumes nearly 13 percent of GNP, which cripples industry. Medical waste is emerging as the largest single hazardous disposal headache. Worst of all, many cultural dilemmas are intrinsic to industrial medicine itself. Socialized medicine, as it is manifest in Scandinavia, fairly delivers services and preventive care, but environmental costs are similar.

Industrial medicine is a downstream solution and is fraught with serious ethical questions. It has become so expensive that the upstream solution of prevention has finally regained popularity. The basic idea is that because getting sick is so expensive, the better alternative is to stay healthy.

Wearing seatbelts is an example of an upstream industrial approach to a serious medical need. The general drop in automotive fatalities in North America can be attributed, in large part, to better trauma treatment. Huge applications of industrial medicine have saved the lives of many who would have otherwise died. Often, however, the lives that are saved are not very fulfilling because of permanent injuries—many of which are to the brain and nervous system. Most head injuries in automobile accidents are the result of the head striking some portion of the interior of the car. In the past 15 years, improvements in automobile design have greatly reduced the possibility of injury. There is a catch. None of these design improvements work unless people take the time to buckle in. Seatbelts are a public health issue. Even with airbags, seatbelts must be worn.

Passing mandatory seatbelt legislation was so difficult that it does not bode well for other forms of necessary upstream public health changes. The rise of neonatology, the medical practice of treating premature infants, is illustrative of the medical infatuation with downstream solutions.

Neonatology has two general forms of patients: those with low birth weights, and those with birth defects. Low birth weights often associated with premature birth have obvious upstream solutions. Prenatal care for the mother, mostly in the form of proper nutrition, could solve the medical tragedy of low birth weight. This solution is unpopular for political reasons. The medical lobby has far more clout than poor mothers so while governments willingly pick up the high cost of treating the prematurely born, much more cost-effective nutrition programs are cut. As it can easily cost $500,000 (1992) to treat a premie, and the treatment itself can cause expensive lifelong medical troubles such as brain damage, even expensive nutrition programs become exceedingly inexpensive by comparison.

The upstream solution to birth defects must address another problem. It can be argued that most birth defects are environmentally caused. It makes intrinsic sense to so argue. The theory of evolution teaches that species mutate in response to a changed environment. It makes perfect sense to assume that a child born without a skull, or some other ghastly birth defect, is simply nature's experiment in trying to produce a human who can cope with PCBs in the water supply or some other form of toxic pollution. The rate of children born without significant defect has fallen to 88 out of 100 in the United States. The effect of toxins in the environment on the birth of healthy children is an emerging health issue. Getting rid of toxic waste is the upstream medical solution. If a significant fraction of the resources given over to industrial medicine were applied to toxic waste disposal, the issues might be resolved. The first step is to connect the two.

Population control is the ultimate upstream-downstream issue. No matter how many elegant solutions are found to the hazards of linear industrialization, the planet is still finite. Linear industrialization could continue in its present realization for a very long time if the populations were small and stable. Almost any practice known to humans is harmful if enough people do it. High populations put intense pressure on the biosphere.

Because the biosphere is finite, any form of human industrial life-support activities must also be finite. Those who would argue against population control or believe that the solutions will take care of themselves, should be required to explain how the planet is going to support these new humans or be ignored.

ELEGANT TOOLS

ELEGANT TECHNOLOGY ASSUMES the existence of elegant tools. Tools, according to Bronowski, drive the development of the human species. Peaceful industrial cultures arise when a society places more importance on the development of tools than on weapon improvements.

Tool-driven social and cultural advances are not automatic and there is often a large time lag between a major tool improvement and the social response. The tool that drove the Protestant Reformation in Germany—the printing press, was perfected over a half century before the social change.

Those not familiar with tools may even recognize this time lag in the history of music. Most musical forms have a classical period. At some time following the perfection of a new instrument, there is a period when the music written for this instrument is considered the best of the breed. For the pipe organ, the favorite composer is Bach: the piano, Beethoven, Chopin, and Liszt: the large concert hall-large orchestra, Beethoven: the small concert hall, Hayden and Mozart: electric guitar, Chuck Berry, James Brown, the Beatles, and the Doors.

All have become standards of excellence by which all later efforts are compared. Subsequent composers are considered derivative. Orchestra boards attempt to educate their patrons with new music while everyone wants to pay to hear the old warhorses. Rock music has spawned radio stations devoted to playing classic rock exclusively. Popular music written from 1964 to 1974 will be played forever—second generation fans of the Doors and the Who exist already.

The pattern of technologically driven change appears constant. In the above examples, the science came first—sufficient precision to manage compressed air in the case of the pipe organ. In its day, the tracore-action organ was the most sophisticated and difficult construction feat attempted by humans—the sixteenth century's space shot. A cast-iron frame gave the piano its large sound and dynamic range. Improvements in stringed instruments and concert hall construction made possible the glories of collectivized sound. Beethoven has been called history's greatest acoustical engineer. Electric amplification made it possible for a tiny group of self-taught musicians to play their music for the whole world.

The time lag between the appearance of new technology and the classical period of the new art form seems about 30 years. The generation of people who produce the new technology seems less creative with the possibilities than the following generations that assume the technology's existence from birth. It is then the creative geniuses appear who exploit the new technology's most fascinating possibilities. In the case of music, a finite number of melodies are pleasing to the ear. Discordant orchestral music causes accelerated hearing loss as the Concertgebouw orchestra of Amsterdam recently discovered. The classical period in music occurs when a composer, or a handful of composers, use up all the pleasant melodies of the new technology.

Since the industrial revolution, producers have been forced to choose between inexpensive and versatile tools that produce primitive products, or expensive precision tools that are the requirements of technical excellence. Precision is expensive: How precise do you want to be?

The tool dilemma of industrialism has been solved. In a technological *tour de force*, Toyota stunned the industrial world with the introduction of the Lexus 400SC. Car and Driver called it the automotive equivalent of "piling on" in football. It was Toyota's way of telling the automotive world—whatever you make, we can make it better.

Toyota has no magic pipeline to superior tools. Therefore Toyota's message of the 400SC is even more telling—*it is not the tools but how the tools are used.* More telling yet—Toyota's most productive engine plant uses tools at least 30 years old. A plant engineer explained that this time was necessary fully to understand these classic machines.

Normally, when an automobile maker brings out a sedan designed to be completely new, the sport coupe derivation is considered a marketing device sold to a small group of aficionados. In essence, the price of a coupe is higher than a sedan with two more doors.

Toyota changed the rules. Their sport coup would cost less and would use over 80 percent new parts. These were not just any new parts, either. The SC was styled in Southern California by artists who were instructed not to use pencils.

For much of the history of machine tools, precision was a function of straight lines and circles. Other shapes cost more to produce and were made with less precision. Using computer-operated tools brought precision and equal cost to the manufacture of any shape. The instruction to the stylists—think shape, not line. The stylists dreamed up the shape of the 400SC by hand forming wet plaster in balloons. It was not only the overall shape of the car that was sculpted, all the parts of this new car were sculpted—down to the last knob and speaker grill.

Toyota showed true commitment to their promise—very few of these delightful sensual sculptures—designed to be touched—were overruled for production or cost considerations. Such a commitment to artistic purity is still rare, but proves an important point. Toyota demonstrated that they could fabricate any shape from any material for any reason. They proved that the problem of production is lack of imagination because *everything* is possible.

Though an enlightened tool-driven culture is *possible*, it is not *inevitable* that tools will lead to the promised land—*there is no technological determinism.* A successful tool-driven culture reflects a *conscious social choice.*

General Motors has tools that make Toyota's look positively primitive. Former GM CEO Roger Smith spent $20 billion in one decade for better tools. He could have been a producer hero-but he will never be. Smith, a living advertisement for outlawing accountants as heads of industrial companies, did not know why he was buying these tools. The only legitimate reason for new tools, during Smith's crazy reign, was job elimination. GM market share plunged from over 50 percent to less than 30 percent. The average age of the buyer of GM's products rose to between 55 and dead. Job losses destroyed GM factory towns.

Industrial environmentalism is the ultimate possibility—made possible by elegant tools. Again, this is not technological determinism—cultures must choose to use the new tools for this end. Evidence is still slim that elegant tools lead to elegant production. Fortunately, there exists a perfect example of environmental choice driving the uses of technological possibilities.

Sweden decided by election to abandon nuclear power. This was a grave choice! Nuclear power drives Sweden's high living standard. There are no simple alternatives. Solar power is weak at her far northern latitude—Stockholm's days are less than two hours long at the winter solstice. She has no oil and the few remaining supplies of metallurgical coal are too precious to burn—besides global warming is a hazard and burning coal is not an environmental improvement. Good hydropower sites have been developed.

One of Sweden's hero-occupations is mechanical engineering. If a replacement for nuclear power was to be found, it would be an engineered solution. Because space and water heating was the largest single source of energy consumption, conservation here would produce the largest energy savings. A house that used less energy was the socially defined goal. Ideas that would lead to no energy consumption *at all* would be entertained.

This effort in cultural ways resembled the American space program of the 1960s. Large corporations and universities joined forces. The first major decision was to produce an airtight, super-insulated house. It was assumed that such a house could not be built unless modern factory methods were employed. To be airtight and energy-efficient, the house must fit together as well as a Volvo's engine.

Folks do not like factory-built housing because of its well-deserved reputation for ugliness. Borrowing a winsome idea from the Danes, the Swedes decided that precision-crafted manufactured parts of a house could be assembled like a giant Lego® toy. These super-Lego parts could be as large and complicated as necessary so long as it could be trucked down the highway. Each fresh house design would be analyzed by computer and broken into to super-Lego components. Anything from a mother-in-law's cottage to a mansion could be built with these methods. Anything, as the Toyota's stylists were told, would be possible. If *these* factory-built houses were ugly, it would not be the fault of the factory.

Computers would direct production machinery which meant that a materials list, down to the last screw, could be calculated instantly. A new home-owner could be informed of each modification's cost with certainty—not only in Kroner cost but in energy consumption. With factory methods, the energy performance of a new house could be projected as accurately as the mileage on a new car.

Because factory housing was an environmental project, environmental design was given new openings. It was discovered that a program could generate a landscape planting design based on each building site's unique environmental setting. By analyzing the path of the sun and the prevailing winds by day and speed, this program will specify the best tree species and location to block the wind, shade the dwelling in summer and allow the sun to shine through in winter. This program can recommend window locations, size of eave overhang, and door location.

Factory housing had satisfied the architects, the consumer advocates, and the environmentalists, but all objections were still not overcome. The health experts were concerned about the indoor air quality of a super-tight dwelling. Moisture and radon gas were big sticking points. High moisture grows molds and fungus which can cause health problems and structural damage. Elegant design solved this concern. An exhaust system drawing from the bathrooms, laundry room facilities, and the kitchen would use an air-to-air exchanger to warm the air drawn in from the outside. By equalizing air pressure between indoors and outdoors, this exhaust system saved far more energy than it consumed. Indoor air pollution was eliminated.

These houses perform magnificently. In many examples, body occupant heat and electric lighting are sufficient to heat the home on most days—even in winter. In one culturally telling example, twelve homes are provided with hot water and indoor warmth from the exhausted heat of the refrigeration unit for a hockey rink.

As every producer knows, great projects only succeed if they are funded. Banks had to be convinced that this method of home-building was a good investment. High-performance housing was more expensive. Banks had to be shown that the energy savings paid for the better house. Because factory methods are so reliable, the energy consumption of the new home could be predicted for 50 years—longer than a mortgage. The sophisticated argument postulated that because this new housing was less vulnerable to an energy price increase, the owner was a more reliable source of mortgage payments and deserved an interest-rate cut. Lower interest rates changed the energy equation so that more sophisticated improvements could be designed into the new house without raising the monthly payments. By designing energy efficiency into the structure and landscape, these improvements became permanent.

Because these houses are somewhat portable, they can be erected at great distance from the factories. When England expanded its military presence in the Falkland-Malvenas Islands in 1982, it faced a severe housing shortage. Within weeks, a Swedish factory was producing the parts for cozy, energy-efficient housing for a cold, windswept island near Antarctica.

Before the dollar devaluation of 1985, Sweden even tried to sell this housing in the cold-weather parts of the United States and Canada with some success. It was discovered that if the military is not paying the shipping bill, the practical economic limit is about 50 miles. The Japanese, world masters at spotting a good industrial trend, duplicated the production facilities and wrote new environmental programs.

One can only dream of how the world would be different, in terms of global policy and internal disorder, if the United States and the old Soviet Union would have employed these methods. Tragically, both had all the necessary tools.

Tools are only tools. They are promise, not results. The culture has to decide how they will be employed. Industrial environmentalism is only possible—not inevitable.

The essential element of elegant technology is design excellence. Design excellence is a product of environmental awareness. Tool sophistication is an important element of the industrial environment. Truly elegant solutions to industrial-environmental dilemmas include an awareness of the on-site access to various levels of technology. Elegant design must take this into account.

Because elegant tools make mass production unnecessary, the problems of mass-production, from the wastefulness of excess capacity to environmental ruin, are no longer imperatives of the tools themselves. Many of the design compromises necessary for mass production can be no longer justified. This leads to a significant shift in thinking. There is a big difference between designing the perfect housing unit and designing the perfect house for a particular site. As each site is different, so each housing unit should be different even if the same design criteria are used. The example of housing is illuminating because people are already aware that different sites require different structures.

The unimaginative who fear social change see the computerization of tools as a capitalist plot that will be used to enable higher rates of production with fewer workers. Often, they have a strong point because the equally unimaginative managers of industry have mostly confined themselves to this application. In this case, the fault is not with the tools, but with the misunderstanding of their potential.

The industrial environmental solution will not fail for lack of tooling potential. In fact, all the pieces may already be in place. The last missing piece in the tooling puzzle was the necessity that computer chips were formerly mass-produced items. Computer chip manufacturers were about as capital intensive as steelmakers. By 1991, the chipmaking tooling has become so sophisticated that designer chips can be made in very small numbers. Chipmaking may become a cottage industry, which in turn means that micro-fine, computerized manufacturing tolerances are available to production runs of ONE.

If tools are no longer the restraint, then the problems are social and economic. The end to mass production, mass consumption, and mass distribution will require major cultural changes even though the alternative is so promising. This is true especially in the United States where size is considered the ultimate measure of the success of any endeavor. The universal availability of very sophisticated small-scale manufacture may lead to the desired regionalization of technology. In response to the industrial environmental needs, each locale has a slightly different subset so each solution must be different. It is comforting to know that tools are not the issue. It is distressing to know that the dilemma is now merely lack of imagination.

Social institutions do have a way of responding, however slowly. If the cooperative was the social invention to respond to mass-production, the franchise is the response to the demassification of industry. What is relevant about the franchise idea is that it is a way to spread industrial recipes in a commercially valid way so that consistency can be maintained in essential areas. As the franchise idea has grown, the distinction between what is essential and what is not has become more sophisticated. In the early days, for example, fast-food franchisers insisted that the buildings must be identical. Now, individual operators have wide latitude over architecture, decor, suppliers, community involvement, hiring practices, and so on. Even so, a Big Mac tastes pretty much the same in Tokyo or in Memphis so tight control is maintained over the basic recipe.

Whether franchising will come to the business of recycling waste control filter elements or some other form of Industrial environmental necessity is an open question. What is known is that the franchising idea could become much more sophisticated in scope and application. Preparing food is one of the simplest industrial recipes. There is nothing to prevent a complex recipe from being distributed through the same social device.

If the system of franchising is not up to the task of industrial regionalization and demassification, then a new social and economic mechanism must be invented. What it will look like or be named is not known. What is known is what it should do. The trick is to invent a social-economic-distribution system that will enable more expensive, well made and sophisticated products to be sold more cheaply. For example, when a product must sell for four times production cost to pay for distribution, there is a real incentive to lower production costs below the point where a proper product can be produced. There must also be new tooling arrangements to prevent overproduction. The waste of overproduction is no longer environmentally, socially (as in jobs lost) and economically affordable. The goal is to make the minimum number of ideal products rather than the maximum amount of junk.

Elegant Energy Applications

NO INDUSTRIAL-ENVIRONMENTAL SOLUTION is possible without a discussion of energy. The second law of thermodynamics cannot be ignored. Organic fossil-based energy use is the prefect example of a linear system built into the industrial infrastructure. Waste steel can be collected, toxic waste can be traced and treated, but it is impossible to collect the energy of the urban heat island. Once energy becomes diffuse, it lacks practical value. Unfortunately, the one-way path of energy is the paradigm of the industrial state. Run short of high-grade energy and linear industrialization stops. Fortunately, high-grade energy sources are not necessary for most of the uses to which they have been applied.

The key to elegant energy application is to use little, employ many renewable sources, and reuse as much high-grade energy as possible. The industrial-environmental solution to linear energy must be a combination of conservation, renewability, and reuse. All must be employed because energy is the power for industrialization in all its forms.

In the United States, approximately 50 percent of energy consumed goes for space heating and domestic hot water, 25 percent goes for electrical generation, 20 percent is used for transportation, and 5 percent is used for manufacturing and industrial applications. Each use could be reduced, in some cases dramatically. The use of high-grade energy for such low-grade applications as space and water heating is likely to be affected by conservation efforts but all forms of energy use can be reduced through available technical means. What is important is that energy use is a function of design. It is built into the system. The United States' consumption patterns are a function of the nature of the industrial infrastructure: how cities are laid out, how food is processed and distributed, how efficiently fuels are changed into useful work, and how buildings are insulated.

Generally speaking, manufacturing processes and transportation require high-grade heat energy, space heating requires a very low-grade form of heat, and electrical generation falls somewhere in between. High-grade sources of energy are finite and increasingly rare. Lowering consumption of high-grade energy, a laudable and necessary goal, must be addressed by technical improvements: making a car go farther on a gallon of gasoline and by structural changes, such as converting a low-grade need to use a low-grade source of energy. Most people have some experience with technical improvements but do not understand structural improvements.

Most electricity is generated by steam turbines. Water is heated in a pressurized system that forces a super-heated steam through turbine blades. The electricity generated is about 40 percent of the heat used to heat the water. This may seem like a low rate of conversion efficiency but it is quite good. Gasoline powered automobiles are about 20 percent efficient, even though the car may get very good gas mileage. In cars, the excess heat is removed through the exhaust pipe and the radiator. The heat lost is low-grade heat in that it is no longer useful for powering an automobile. In electrical generation, the low-grade heat is dispersed through cooling towers and warm water dumped into rivers where it enters the environment as thermal pollution.

The exhausted low-grade heat from electrical generation still contains useful heat energy. In most countries of northern Europe, this water is piped through cities and used for space heating and domestic hot water. This system is usually called district heating. With district heating, citizens get electricity and heat for the same amount of energy as formerly only got them electricity. Forty percent of the energy still goes for electrical generation and the 60 percent is not discarded but is used to heat buildings and water.

District heating is an extremely good idea. It does not solve the linear direction that energy must take to conform to the second law of thermodynamics; it merely reuses the energy more than once as it tends toward entropy. Structurally, it is an elegant solution. Once the pipes have been laid and the homes converted to district heating requirements, any form of low-grade heat can be used. Where the buildings to be heated are very weather-tight to begin with, very small sources of low-grade heat can go a long way. Burning refuse can yield home heat with district heating. Most cities in the United States north of the 40th parallel could profit from the installation of district heating.

District heating is not the only example of changing the quality of the heat input. Electric trains are another example. Diesel fuel, which is used by most locomotives in the United States, is a high-grade heat source. It is also a premium fuel that is rare. Power a train with electricity and you can power the train with garbage, coal, falling water, wind, peat, and other sources of energy more common than diesel fuel. Since electricity is easily transported, the source of power does not have to be near the train. It can be located close to the energy source or close to a city where the waste heat can be used.

Combine technical improvements with structural improvements and only manufacturing, cars, trucks, airplanes, and farm machinery require high-grade sources of fuel. Low-grade sources of fuel are sufficient for all the other applications. Make the energy requirements low enough and the world of solar power become a distinct possibility. Solar power is plentiful but diffuse. This causes a technical headache when solar power is to replace conventional fuels. Attempts have been made to fashion a solar powered airplane or automobile without much success. A solar powered aircraft has flown over the English Channel but it was an experimental rather than a serious form of alternative transportation. Flying is a high-grade energy proposition. No meaningful discussion of the future can exclude the subject of solar power. For the immediate future, however, solar power must be confined to low-grade applications.

Solar power encompasses technologies such as biomass, wind power, solar concentrators, and photovoltaics. In the early, optimistic, and experimental phase of solar power, it was assumed that putting a solar collector on the roof would lead to everything from power-sharing to sinless perfection. Simple solar collectors did not work, unfortunately. The notion that a portion of the roof could collect enough energy to heat the interior of a conventional house was technologically preposterous. This phase was referred to as "active" solar power. Almost no examples of active solar worked. What proved to be much more successful is something called "passive solar." The idea of passive solar is to make the building as energy efficient as possible with insulation, quality windows, and an air-tight weather envelope. When the structure has energy needs so low that a human body or its lights can keep rooms warm under most conditions, then a proper orientation to the sun, well-designed overhangs, and solar porches become enough to heat the home. The only active solar collector that has proven itself is one used for hot water. This is logical because the dwelling only has to be warmed to 70° Fahrenheit whereas water must be heated to about 110° Fahrenheit.

Solar power as wind power has a bright future. The whole of the United States is linked together by an electrical grid. Where the power is generated is largely irrelevant and some areas of the United States have a plentiful supply of wind. North Dakota is a very unpleasant place to live and not many people have made it their home—yet it sometimes called the "Saudi Arabia" of wind energy. One of the worst features is the wind that blows almost constantly. People who live in North Dakota look on their windswept prairie as an ocean where exposure to the elements is incessant. Only the towns offer protection from the wind. A string of windmills along the North Dakota-Canadian border could produce electricity about 80 percent of the time which is about double the operating record of a nuclear power plant.

The key to wind power is embodied in the notion of a string of windmills. Experiments in wind power have concentrated their efforts on large windmills. This is contrary to good industrial practices. In leisure-class parlance, mass production is associated with shoddy production. This is false, of course, because the more examples of a single design produced, the more time, effort, and money can be spent on perfecting the design. Reliable windmills mean reliable bearings, lubrication systems that function in all weather conditions, blades that are not damaged by rain, sleet and snow, easy maintenance, and a whole host of considerations that will not be addressed by large, one-off, prototype windmills. Only mass-production techniques will make wind power a viable energy alternative. Many perfected windmills are preferable to fewer large windmills for another reason: when one fails, the results are much less noticeable. As for the social realities of mass wind power, North Dakotans already put up with Minuteman missile silos, so a string of windmills should cause little concern.

THE ELEGANT MIX

THE INDUSTRIAL-ENVIRONMENTAL FUTURE becomes most attractive when all the elegant methods are combined. Imagine a city with a resource recovery center that burns flammable waste for electrical generation and space heating, composts and concentrates organic waste for use by agriculture, and collects and recycles metals for reuse including the building of other recovery centers, laying the pipe for district heating, and infrastructure improvements. Imagine a city that reprocesses all toxic waste. Imagine a city without operating landfills and smokestacks. Imagine medicine designed to keep people healthy and when the time comes, lets them die with dignity. Imagine a system planned so that each new member of society has a place to fit in an environmental and industrial sense. Imagine a society where agricultural practices can continue far into the future without causing devastating environmental consequences. Imagine all that, and the possibilities of industrial environmentalism become real—a pleasant thought.

The scale of industrial environmentalism is at the same time its greatest attraction and its most serious drawback. It is a huge scheme. Any responsible person must ask how such a scheme can be politically and socially possible, and more importantly, how it will be paid for. These questions are not frivolous and the answers will constitute the remainder of this book. The question of what may happen if industrial environmentalism is not tried will also be addressed.

CHAPTER TWELVE
BUILDING A CONSENSUS FOR ACTION

THERE HAS BEEN NO SHORTAGE of suggestions for rebuilding the industrial states—mostly because anyone who believes that everything is fine belongs in a fool's paradise.

The failure of most suggestions is that they neglect the nature of the producing classes. E. F. Schumacher wrote a wonderful book called *Small is Beautiful.* His descriptions of the industrial dilemmas were magnificently stated. When it came time to make suggestions for improvements, he called them "intermediate technology." If there ever were a notion unlikely to be embraced by the producing classes, it is something called intermediate. Producers are likely to be as thrilled with an "intermediate" future as getting a mediocre grade of C- in a college calculus course. It is not their nature.

Social progressives who believe that producers will be enchanted with an intermediate technology agenda are as likely to be disappointed as the Marxists who tried to recruit farmers and trade unionists to their agenda. The fact remains that the producers believe, rightly or wrongly, that they can cut a better deal on their own without the help of such technologically illiterate progressives. In the main, social progressives misunderstand the people who create technology as profoundly as they misunderstand technology itself. They look at the poor migrant farm workers, the miners, or people who work on assembly lines and conclude that a radical social prescription is needed dramatically to improve lives at the bottom of the social heap.

Not all producers, in spite of the historical record, are found at the bottom of the heap. Within the range of what they want to do, many producers find themselves at the top of a very interesting pile. Marxist class analysis has little to say about the person who designs computer software while sitting in a hot tub. While the social progressives like to prattle on about the wonders of social equality, the producers shut them out. Equality, for them, would be a large step down. Equality is a benign idea that has enough universal appeal that even the producers who do not believe in it will not object in any strenuous way. They view equality as Jeffersonian, that is to say that, in a social context, producers should be equal to any predator, but as was pointed out in earlier chapters, they do not even believe this very much.

There is a concept that does really raise the ire of the producers. This is the idea that technology is dehumanizing. Say that technology has somehow diminished the human condition and the producers will tune out the person who raised it. This is not surprising because this statement is nothing short of a vicious insult to the work of the producers. It is also demonstratively false. Although factory work may be dangerous and boring, the products of factories have made a change for the better in the human condition. Moreover, factories, or any other large-scale projects, are not the products of the efforts of Martians—they are the products of Earthlings. How advanced technology can come to be called dehumanizing is an issue that only a social progressive seems able to understand.

Because the industrial state is in such serious trouble and because the producers themselves find it difficult to understand the social implications of their work, there is a need for outside social guidance. Outside guidance is likely to fall on deaf ears, however, unless it considers the reality of the producer elites. Such elites can be seduced into an alternative industrial agenda but as with any other seduction, the interests of the seduced cannot be ignored.

Unless outside guidance is really exciting, the temptation of the producers is to fall back on an age-old response called the "not-invented-here" syndrome. Inventors with new ideas, who approach large companies with in-house research and development staffs can describe this syndrome well. The inventor's idea, no matter how good, encounters stiff resistance because the company management believes, with some justification, that if the idea was so good, why did not their expensive in-house research staffs come up with it themselves? If producers act this way with the ideas of other producers, imagine how they respond to people who are not producers.

In order to seduce the producer elites with an alternative industrial agenda, the agenda must be exciting enough to get their juices flowing. The idea of "intermediate technology" will *not* do the trick. One of the reasons solar power has not attracted attention is largely related to the way it has been described. The early solar power advocates talked about its simplicity. Of course, harnessing solar power is neither simple nor cheap, but because solar power became associated with a technology appropriate for underdeveloped nations or spaced-out hippies, it immediately turned off the industrial elites who want their projects large and complicated, and *perceived as such.*

Interestingly, the advocates of solar power have come to believe that the solar agenda is in trouble because the utility companies cannot put a meter on the sun. This is preposterous. The utility companies cannot put a meter on coal or uranium ore either. Where they put the meter is between the technology and the user. Solar power will not substantially change this equation. It is unlikely that the solar advocates will ever understand that their dreams have died precisely because of the way they sold their agenda. It simply was not complicated enough to attract the attention of the very people necessary to make it work.

Calling the new industrial agenda elegant is necessary, but hardly sufficient. To attract the attention of the producer class, the agenda must be large enough to employ significant numbers of the producer classes for a lifetime. What, in fact, will fire the imaginations of the producer classes is not elegant technologies, but elegant mega-projects.

BUILDING FOR THE FUTURE— INDUSTRIAL ENVIRONMENTALISM AS A DEVELOPMENT STRATEGY

THE CRISES OF INDUSTRIALIZATION affects everyone. There are no known escape routes from the industrial crises without collective action. The problems are caused by collective action; there are no innocent parties in industrial societies. Only successful businesses become environmental problems. Failed enterprises cause no environmental burdens, as a rule. This means that people who blame companies for environmental problems are really blaming themselves. A poor old person huddling around an electric heater in a cold leaky house would not seem to be a villain, but the power used and wasted because the home is poorly built causes acid rain or nuclear waste. The old person is not an ecobandit of the stature of a James Watt but is still far from harmless .

Collective action is also necessary because everyone has a stake in the outcome. Some might argue that there is no hope for collective solutions, especially in countries such as the United States where individual action is so highly venerated. The solution, according to such people, is for everyone to clean up his or her own act. If everyone would cease his or her environmentally dangerous practices, the stresses would end. Buy a plot of land in the wilderness and build an environmentally sound structure and at least this one individual would cease to be a part of the environmental disaster.

This solution is nothing but a fantasy. Assume that the environmentally concerned person chose to follow this path. Further assume that this person can build anywhere, has no need for employment because of independent wealth, and has a large amount of cash to build the perfect place where environmental problems will not intrude and there is perfect harmony with the surroundings. The first question is purity: are manufactured pieces going to be used to construct this Nirvana? Can an environmentally sound dwelling be built without glass, nails, insulation, caulks and sealants, roofing materials, surfaced lumber, mortar, adhesives, and a litany of plastics? All of these cause pollution during manufacture so even the most well laid plans are going to yield pollutants in the construction phase. With a great deal of money a small-scale composter and waste treatment system could be built that would minimize ground water contamination. By reducing food variety, a small-scale agriculture could be organized so that a low level of self-sufficiency could be achieved, but there are no easy sources of power and none whatever for medicine, and there is no escape from the pollution that comes from outside.

No area on the planet is immune to acid rain. The individual who moves to the wilderness to escape the city and industrialization finds escape from air and water pollution impossible. The individual has not escaped the life-sustaining forms of industrialization (nails, farm implements, medicine) nor the death dealing realities of industrialization (acid rain, poisoned groundwater) The individual cannot escape the responsibility the problems themselves. Five thousand years of back-to-nature fantasies have only proved that Eden is not possible.

A fascination with individualism is not the only impediment to collective action. Even countries such as the United States can agree on collective action when there is a sense of urgency—ask anyone who lived through World War II. The problem with the industrial environmental crises is that the changes are so slow that the urgency born of crises cannot be mustered. Even a subject as peripheral to the issue as the nuclear arms race gets more attention than the central issue. Nuclear weapons at least threaten the possibility for a dramatic event. Even if the issues alter slightly in the nuclear debate because of changes in technology, the possibility of a life-ending bang remains.

For, whatever the dangers of nuclear weapons, until the missiles are launched, there exists the possibility that they will not be used. Nations have the option to dismantle these weapons or insure that they are never used. Given the human fascination with war and loud noises, this possibility may seem remote, but is still a possibility. In a very real sense, this possibility no longer exists for the slow-change problems of the industrial environmental crises. When it comes acid rain and groundwater contamination, the missiles have already been launched. The poisons that will affect groundwater are already in the ground; the forests have already started to die. The industrial environmental problems will snuff out human life as certainly as war. Death by vaporization in a nuclear conflict will be swift. Death by pollution or starvation is slow by comparison. The slow-change nature of the industrial environmental crises has diverted attention away from its serious nature. Lack of attention diminishes neither the seriousness nor the urgency, it merely delays solutions. If dramatic issues such as the nuclear arms race have defied collective solution, imagine the lack of reaction to a slow-change issue.

MEGA-PROJECTS

THE GOOD NEWS about the industrial environmental crisis is that it is technical. The bad news, especially in the United States, is that technical-industrial topics are beyond the scope of intelligent public discussion. Industrial environmentalism is an industrial solution and has no natural political enemies. Industrial cooperation is more common than political cooperation. Even in that den of bickering, the United Nations, organizations in charge of technical issues such as meteorology, international travel, and communications are quite successful. The only thing standing in the way of an industrial environmental solution is lack of awareness of the problems and potential solutions. Ignorance and apathy are enemies of collective action but in the absence of organized political resistance, such obstacles can be overcome.

Interestingly, the social and economic realities of the industrial states in the 1990s are fertile ground for an industrial environmentalism. The world's industrial economies are in sorry shape. There is widespread unemployment—especially among the industrial classes, there are unfunded liabilities, most of the industrial infrastructure of the United States is obsolete, and economic growth, by any common definition, is impossible because of resource limitations—especially in the area of energy.

These may appear to be horrendous constraints, but they are not. Unemployment means there is a significant pool of skilled labor. Debt and unfunded liabilities mean that there is a crying need for productive investment. Replacing an obsolete industrial infrastructure is by definition a productive investment if the new infrastructure solves the problems of the old. Closed-loop industrialization *will* solve the problems of the old industrial infrastructure. Properly done, economic development in the industrial-environmental mold will solve these problems simultaneously. Closed-loop industrial solutions require closed-loop problem-solving methods.

Industrial economic activity can be likened to throwing a stone in the water. From the main project there are economic ripples. Put up a building and the construction workers spend their paychecks on food, clothing, and services. The construction project is the stone in the economic lake. The services which spin off from the initial economic splash are the ripples. If the economic rock is big enough, the ripples are profound. The construction of a dam can caused the formation of a small city complete with lawyers, merchants, teachers, homebuilders, clergymen, and prostitutes. Disney World, built in central Florida, caused a building boom in hotels, campsites, restaurants, homes, hospitals, fire stations, and the homes for all the people who built and staffed these structures. Disney World was a mega-splash in the central Florida economy that was based on ranching and citrus farming.

The biggest civilian economic rock in the post World War II American economy, however, was the interstate highway system. The interstate highway system produced a building boom that left everything from large suburbs to automobile factories in its wake from coast to coast. Highway building had it drawbacks: farmland lost to production; urban sprawl, petroleum depletion, and environmental damage. In this case, the issue is not the result but the process. The interstate highway system may have been a terrible idea, but its economic impact cannot be overstated. It was a very big economic rock.

Disney World may be considered a small mega-project; the interstate highway system was a *mega-project!* The economic problems facing the industrial states are so large that it may require many mega-projects to restart the industrial engine. When selling industrial-environmental mega-projects in the United States, it may be instructive to remember how the interstate highway system was sold to the American public. When the interstate highway system was being considered, the economics of development were discussed but briefly. No one imagined urban sprawl and a country with 150 million vehicles. Those who made construction equipment imagined the trade in roadbuilding machinery would improve but few imagined the impact on the home power tool market with all those thousands of suburban basements and home-improvement projects. Trade unions supported highway building because it meant jobs. No one foresaw the impact highway building would have on the plumbing business.

Initial support for the interstate highway system was weak even though it was widespread. What made the interstate highway system a winning proposition was when a President, who was also a war hero, began calling these highways escape routes to be used in time of war. These were not merely highways, claimed Eisenhower, these were a necessity for national security. Once the arms race reached the stage where leaving the city was a ridiculous idea and the constituency that directly benefitted from the highway system grew, the defense argument for the highway system was quietly dropped. Few Americans alive at the time of the original decision to build the interstate highway system can remember the defense argument. It is sad that politicians will vote for a defense issue where they will ignore an economic development one. Such is the nature of leisure class politicians, virtually the only type of politician the United States has. The issue is not to argue with but to work with this reality.

If the interstate highway system can be sold on the basis of national security, the mega-projects of industrial environmentalism can be sold in like manner. It is time to make people aware that the slow-change threats to their life are as real as the threats of war. It is a more urgent than military conflict because the war with the environment is already raging. It is a war that the environment will win. Beyond the obvious decision to market industrial-environmental solutions as issues of defense, there is the problem of proper mega-project packaging.

Forging a political and social consensus for industrial-environmental mega-projects is a matter of uniting the various groups for whose self-interest industrial-environmental work. The most obvious interest group is the young. Slow-change problems are more meaningful to a person who has 60 years of normal life expectancy left than one who has 10. Threats that the world will run catastrophically run short of fossil fuels by the year 2020 are more real to a person who will just be 40 than to a person who expects to have been dead for 20 years. This is not to say that there are not environmentally concerned elderly persons. Age brings the wisdom of the concept of limits. Applied to the subject of the environment, such wisdom in the old can be quite profound, but such people are saints and the normal tendency is to be quite unconcerned about the condition of the world following one's death. Sainthood is relatively rare and should never be confused with politics.

In truth, the future, to even the most modestly aware young person, must look very bleak. The planet is crowded, the water has been poisoned, the soil has been eroded, the economic base is shrinking, the factories and other infrastructure are worn out or closed, and the prime natural resources, from sweet crude oil to 70 percent iron ore, have been taken from the earth and turned into a large junk heap. If this were not enough, government has incurred huge liabilities in the pursuit of this stupid plunder and the young will be stuck with a huge monetary debt to go with the huge environmental debt. Only the ignorance of youth has kept this group from rioting, so insanely bad is the future they face.

The mega-projects of an industrial environmental solution will change the future dramatically. The energy and idealism of youth will be channeled in a constructive way. The youth will be building the country they will inhabit. They will feel control over their destinies. They will have useful and necessary employment. They will even feel a deserved sense of arrogance as they unbuild then rebuild the mess left them by their parents. It is likely that any or all of these advantages will be enough to get a young person solidly behind an industrial environmental program.

Natural members of the industrial class would love the notion of mega- projects. Building is industrial immortality; not building is a painful and slow death. Industrial unemployment is world wide problem. Putting these people back to work on almost any project would be supported by the producers of society. Building large projects of national and international significance would bring more than support; it would engender outright enthusiasm.

Another natural group to support industrial environmental solutions, is the committed environmentalists. The environmental movement has lost a great deal of momentum since the halcyon days of the clean-air act. Two fallacies dog the environmental argument: 1) Setting aside park land as a museum for wilderness does not preserve the environment. The spillovers of the city can travel anywhere to destroy the most carefully fenced off areas as is demonstrated in the case of acid rain. 2) To date, the environmental movement has not formed an economic paradigm for environmental action combining with economic development. By addressing the environmental impacts of industrialization in a constructive manner (build a way out of the problem)—the environmental movement would have a development agenda to go with their laudable preservation agenda. Such a combination could be the route to real political power.

The group that should be most interested in industrial environmental mega-projects is the investment community. The number of sound investments is shrinking rapidly. Billions have been spent in the United States for the building of commercial office space. The next office building should not be necessary for at least 50 years. The same could be said for hospitals, clinics, and doctor's offices.

The bloom is off the rose in nonmilitary high technology as the world becomes saturated with installed computer capacity. The market for high-priced housing is so saturated, that the majority of units may never be sold. After ten straight years of losses, United States farmers have borrowed to their net worth and then some. Foreign borrowing has been reduced to debt service as country after country has been subjected to International Monetary Fund (IMF) conditions. There are no real productive loans left to make. Loans and investments of all kinds are at risk in a catastrophic collapse—especially during a time when banking practices are destroying the base for repayment.

In order for there to be orderly industrial growth, real interest rates (nominal rates minus monetary inflation) cannot exceed industrial growth rates. Anything higher is predatory plunder that will slow industrial growth and destroy the means for repayment. After over 12 years, in 1991, of interest rates exceeding industrial growth, such predatory practices have reduced the industrial infrastructure of the United States to shambles. The money numbers on the computer screen appear larger but the practice of shutting down factories in exchange for rearranged electrons has significantly impaired the ability of the United States economy to make loans good. There are few solvent banks left because investment practices have been absurd. These practices may have been a social good in the sense that linear industrialization was already in trouble and predatory investment practices have killed a dying goose. This attitude would be perfectly acceptable if investment was already swinging to a closed-loop industrialization so that new systems would be in place to make good the loans.

The most conservative investment advice is to invest in things that a person would want to keep in case there is no way to get rid of them. When an investor buys a stock, or a painting, or gold, he or she is betting that when the time comes to sell these items, there will be a willing buyer at a higher price. This form of investment is called the Greater Fool's Syndrome (GFS) that postulates that any investment, no matter how little its real value, is an investment because no matter how much was paid, at some future date someone will be foolish enough to pay more. The GFS is present and accounted for when old baseball cards, diamonds, antiques, and Franklin Mint Collectables are referred to as an investment.

If something contributes to people's happiness or overall well being, the investment is always secure no matter the future market conditions. The most obvious example of such an investment would be a wine collection. Old wine has been known to appreciate in monetary terms in the past. Even if the monetary value of the wine goes sour as an investment, at least there is a premium wine collection to drink. It may not solve the problem of a lost investment, but it will go a long way to soothe the pain.

Leaving aside predatory investment practices for a moment, consider that a great deal of saving is done to provide for retirement. At some point in the natural life cycle, people are too old for strenuous work. Frailty also manifests itself in sickness. Providing for the late period in life is responsible. Saving, in and of itself, does not provide for retirement, because money is of no value unless it can buy something that is wanted in the future. Only investment in the systems that will provide retirement needs will make it possible for the money saved to buy anything. If investment schemes damage future industrial support systems and damage the environment, the money a person has accumulated for retirement is meaningless.

Like investing in fine wine, investing in industrial environmental mega-projects has the advantage of safety. No matter the future economic arrangements, if the proper systems for support have been built, the most important criterion of a sane investment strategy has been met. Although it is likely that a future form of closed-loop industrialization will support economic prosperity, this does not matter so long as such an industrial environmental plan provides for the needs and wants of the citizens in the future. A great deal of financial activity is managed by professional pension-funds for the purpose of providing for the needs of retirees. It would be in the best interests of everyone, but especially this group, to invest in future support systems. Profits are important but not as important as building for the future. Responsible pension managers should understand this. Unfortunately, they too have played money games while the industrial foundation of the economy crumbled.

Gathering the forces for industrial mega projects is more than identifying common interests. Youth, environmentalists, the industrial unemployed, and investors constitute a majority large enough to make mega projects politically salable. Selling a political agenda is only a first step. A working plan for industrial environmentalism must be in place or it will die the death of all good intentions. The environmental movement has lobbied long and well for such things as the clean-air act and the superfund for toxic waste cleanup, but the air is just as dangerous and toxic waste is worse. Appropriating money without a plan of action only results in office buildings full of sincere bureaucrats who worry a great deal. The problems must be solved. The time has come for action.

Some elements of an industrial environmentalism are already in place somewhere in the world: the French have extensive experience in garbage-to-energy conversion; the Dutch have knowledge of mass composting; the Germans have figured out a way to burn high-sulfur coal without sulfur dioxide emissions—the waste product is gypsum, a useful building material—and have learned to make steel without coke; the Japanese are experts at recycling steel (this year's Toyota was last year's Buick); the Danes have extensive experience in building wind-powered generators; and the Swedes have made giant strides in packaging, district heating, and energy-efficient homes. In some respects, industrial environmentalism requires only an adaptation of systems that have been tried and perfected elsewhere.

Challenges such as increasing the energy efficiency of buildings ten-fold, electrifying major rail links, installing district heating, and building resource recovery systems employ existing technology. As such, building these mega-projects is essentially a matter of political action. Industrial design solutions are highly contingent on the nature of the problem. A good house for Montana is going to be different from one in Florida even if the design goals are the same. An energy-efficient design for Montana will be directed to keeping the inhabitants warm; an energy efficient design for Florida must keep people cool. Houses where winters are cold but sunny will be different from houses where winters are warmer but cloudy.

Just as natural conditions are different from place to place within the United States, so political conditions differ. Take the example of electrical generation. Some electricity is generated by public corporations, some is generated by cooperatives, and some is generated by municipalities. Who generates the electricity is not so important as how the electricity is generated. Is the supply sufficient, and will it run the television? How the electricity is generated is more economically significant than the ownership. If the electricity is generated by old dams, the electricity is cheaper than if generated by nuclear power. The capital costs of nuclear power illustrate one other point. It really matters little to the investment community if a private company, a cooperative, or an arm of a government agency does the borrowing. For this reason, how investment is channeled to a mega-project is mostly a matter of regional preference. Since the same design goals must be met, the same people must do the designing and building, and the same interest will flow to the same investment funds, the ownership arrangements are irrelevant. Industrial environmentalism is neither a communist nor a capitalist plot. Conventional political arguments must not stand in the way of industrial environmental progress because they are meaningless.

Assume that it is in the industrial interest to see that all the major rail links in the United States are electrified. Convincing the railroads that such a project would be in their best interests may be difficult. If one railroad failed to go along with the project, the whole system would be in jeopardy. Locomotives would have to be switched where the electric supply stopped. If one railroad failed to comply, the others would resist because they might be less profitable. The railroads could be compelled by law to electrify major links but such legislation would be nearly impossible to pass.

The key to the industrial environmental mega-project of electrifying the major United States rail links is the common good. By electrifying trains, precious diesel fuel can be saved for applications like flying and plowing fields; the steel industry would be humming making all the towers to support wires as well as any generating equipment be it windmills, garbage burners, or otherwise; the copper industry would profit from the large orders for overhead wires and windings for electric locomotives. The country will always need to move goods about even if petroleum-based fuels run short. Electric railroads are in the common interest even if some interests are better served.

A possible political solution might be to organize the railroads like the interstate highway system. The government owns and maintains the roads while private industry owns the rolling stock. This is also true for airports and river maintenance. The Army Corps of Engineers dredges the channels but business owns the barges. Municipalities own airports but not airlines. None of these state owned transportation facilities is expected to make a profit. If road taxes pay for maintenance and debt service, they are considered successful. By this analogy, it would make perfect sense for government to own the railbeds while railroad companies owned the trains. There is significant political precedent for such a strategy.

The condition of railroad tracks is in everyone's interest. Many hazardous material are shipped by rail. Most train accidents are caused by flaws in the tracks. Good roadbed maintenance can prevent many environmental catastrophes. Such roadbed maintenance is a function of money and labor. These two are interchangeable to some extent. Labor intensive maintenance practices can substitute for expensive machinery. Political decisions as to this choice are possible. A state that believes that driving spikes by hand would be a good summer youth employment scheme, could opt for the labor-intensive strategy. Those areas that appreciate the automated solution could invest in a tie-replacing machine. Maximum regional flexibility could be achieved on a nationwide scale. How something is accomplished is less important than what is accomplished.

Instead of coaxing railroads to build major electrical links with all new tracks, the relevant government agencies would own and operate the tracks. They are better equipped to run the tracks than private business because they have more technical options. The role of the federal government would be to pass enabling legislation to set performance standards, provide suggested options for meeting these standards, and possibly guarantee the financing to soothe the jittery nerves of investors. The job of building the electrified railbeds would be turned over to regional or state authorities with the admonition to have at it.

The electrified railroad mega-project example illustrates the basic industrial environmental political strategy. The method is to think globally, set goals nationally, plan regionally, and execute locally. The key to success is to make each facet of the overall plan most suitable for local conditions. Local control not only has the political advantage of giving more citizens a say in their destiny (government by consent of the governed), it has significant technical advantages. Building an industrial structure to conform with the realities of the environment means the environment must be well known. No one knows the design and maintenance problems of a given area better than those who have lived there all their lives. It is better for a resident of an area to study solutions where solutions have been tried than to rely on outside experts. It is easier to learn railroad construction than to learn the political, social, and technical problems of getting something accomplished in a given locale. This is especially true with mega-projects that have a multitude of ways to achieve the same end.

All industrial-environmental mega-projects that have technical models can be subject to the think globally, set goals nationally, plan regionally, and execute locally strategy based on an interstate highway system model. The role of the central government is to formulate the plans so this strategy can be implemented. District heating becomes a mega-project when the Federal government decides that all cities larger than 5000 inhabitants North of the 40th parallel should have such systems. Garbage-burn electrical cogeneration systems and composting systems would be good everywhere which makes them perfect industrial-environmental mega-projects.

These are examples of off-the-shelf industrial-environmental solutions. Adapting European technology to American problems is merely the beginning of a very long journey toward a closed-loop industrial society. There are many things we do not know. Many inventions will be required. European technology, as good as some of it is, does not solve all the problems. Mass garbage burning schemes, superior as they are to landfill disposal methods, have not solved air and water pollution problems. District heating, weather-tight structures, and electric railroads are intelligent methods of using energy more wisely and would buy valuable time while long-range solutions to the energy problems are developed, but they are not, in and of themselves, permanent energy solutions.

While off-the-shelf European technology is a fine start toward an industrial environmental future, it is only a start. To close the industrial loop will require research and development, wrong turns, stunning breakthroughs, and failed experiments. In short, inventing the future will have all the twists and turns that are associated with industrial experiments of the past. While this is to be expected, it nevertheless is distressing to realize that not all the parts of the puzzle are in place. This is especially true for the United States that does not seem to have the ability alone to lead the industrial pack. Except for weapons, some electronics technology, advanced medical equipment, and pharmaceuticals, the United States has become a second-rate industrial nation. Not only have American industries failed to invent, develop, manufacture, and perfect the technology just discussed, they have fallen behind in automobiles, consumer electronics, tool making, and much else. To ask American science and industry to leap ahead to solving the problems of industrial environmentalism may be asking a bit much given the disorganized state of American engineering.

It is not that American engineers are poorly educated or stupid, it is a matter of working conditions. Engineering is not a glamor profession in the United States Almost no American could name any engineer, living or dead. The conditions for employment in engineering fields are generally much worse than those for other professions of similar skill levels. Because engineering is task oriented, jobs have beginnings and endings much like the construction business. When the job is finished, engineers can find themselves out of work. Engineers have much higher rates of unemployment and job changing than a profession such as accounting. Engineers are not often found within the mainstream of the business culture and few have reached top management, even in industrial companies. (Ford Motor is an obvious exception.)

Students are not encouraged by the awareness of such working conditions to pursue engineering. Even those who are undaunted face a difficult educational experience. Most engineering courses are extremely rigorous and cannot be faked. In many disciplines, a student can graduate with honors without ever having to learn a thing. Engineering students, on the other hand, take tests that have correct answers. It is very difficult to fake a knowledge of calculus, especially in the third year of training. Because the worth of engineering has been so devalued by Americans, many engineering professors in American universities are foreign-born.

The selection process for engineering students, itself, also contributes to the poor performance of American engineering. Students are encouraged to go into the profession if they can demonstrate an ability in math. Math is an interesting subject in that it is one of those fields, with music and chess, where child prodigies exist. The gift for math is probably genetic—either math makes sense and is easy, or it does not and is difficult. An ability in math, while easily recognized and graded, is not necessarily a good indicator of engineering ability. Engineering is a creative building profession. Creativity is more important than math ability to great engineering. Many math whizzes lack the necessary ability to foresee a project that does not exist. Creativity is probably also a matter of a genetic gift, but it is not the same as a math gift nor can creativity be so easily identified. Even so, engineering schools elevate the importance of math to a level of exclusion. In so doing, graduates are often math wizards without a new idea in their heads. Even if creativity cannot be taught, the broad knowledge necessary for innovative, creative solutions can. If engineering schools were expanding the intellectual horizons of their carefully selected math wizards, more engineering creativity could result. But engineering education has become, over the years, less human, more mathematical, more abstract, and more isolated.

In spite of these impediments to good engineering, many fine young students graduate with the requisite abilities to become great engineers. This number is small, but they do exist. What happens next can only be described as the Great American Tragedy. Because employment offers are the sweetest, the best and the brightest engineering talent goes into weapons design. If the national debt is the economic symptom of military extravagance, industrial decline is the real issue. The United States and the old Soviet Union have the most dangerous and destructive weapons and are often called as superpowers, but they are both behind Europe and Japan in many indexes of industrial competitiveness. Diverting the best engineering talent into weapons design has caused the United States and the old Soviet Union to become muscle-bound. Either could destroy the planet but neither can build a state-of-the-art automobile. Both can force their way to the status of empire but neither has much to offer in the way of economic development technology to their client states. As a result, neither of these belligerent industrial powers is particularly loved and while their power is unquestioned, their influence is waning. In truth, having a great deal of military power does not seem to account for much.

Weapons manufacture is industrial waste. Little military science or engineering has civilian benefits. The example of microchips is cited to refute this position but the fact is that microchips are an extension of transistor development— a civilian project of Bell Labs. Microchips could and would have been produced by nonmilitary research anyhow. Even if the example of microchips is allowed as a useful industrial spin-off of military research, the fact remains that useful industrial goods are produced 100 times more efficiently if the research efforts are directed at the wanted item in the first place. The real issue is whether most defense spending actually buys any defense. If it does not, a very large hole is cut in the defense argument for diverting talent from civilian enterprise.

PAYING THE COSTS

IT IS UNLIKELY that industrial environmental mega-projects and the arms race can be funded simultaneously. Money is not the issue. The amount of money in circulation is a political decision and enough money could be made available to fund both industrial environmental mega-projects and a lovely little arms race. The real problem is with a shortage of engineering talent. With the off-the-shelf projects such as district heating, the technology has been developed sufficiently so that engineering would not be a bottleneck. The same is not true for the uninvented elements of the industrial-environmental solution.

Great engineering creativity will be necessary for all the problems that must be solved. Elegant technology requires elegant and sophisticated engineering talent. Even as engineering schools must be expanded and upgraded, this will not compensate for the loss of the precious talent from the top of the class to useless and wasteful pursuits such as new nuclear weaponry. These people are necessary for the defense of the industrial-environmental umbilical cord to which all in industrial state are attached. They can no longer fritter away their talents on military technology based on absurd arguments.

It is not enough for industrial environmentalism to be called a defense issue. It must get a share of the defense budget so that the requisite talent can be hired. If the old Soviet Union, the declared and implacable enemy always cited whenever a defense appropriations bill goes to Congress, had come over and poisoned the water of Americans, this would have been treated as an act of war. Yet Americans have done a fine job of poisoning their water, often while building the tools of war (check out how many toxic waste sites in the United States can be found near locations of weapons manufacture). It is no different if the evil enemy poisons the water or if the water is poisoned building weapons to scare the enemy—the water is still poisoned. The United States and the old Soviet Union have finally found a common enemy; their absurd arguments in the name of defense and the damage they have caused to their industries and environments. Both need a comprehensive military conversion strategy and now they have one.

CHAPTER THIRTEEN

THE HURDLES TO INDUSTRIAL ENVIRONMENTALISM

IN GENERAL THERE ARE FOUR REGIONS that have reached the phase of mature linear industrialization: Japan and Pacific Rim, Western Europe, The Eastern Bloc, and North America. Any of these areas is ripe for a conversion to a closed-loop, industrial environmental agenda.

The hurdles to a conversion to an industrial-environmental program are of four basic types; HURDLES OF AWARENESS, VALUE HURDLES, POLITICAL HURDLES, AND ECONOMIC HURDLES.

Notice that technical hurdles are not an issue. An industrial environmental solution assumes that though the producers are going to be performing different tasks than those they are currently doing or have done in the near past, these tasks are similar enough to current efforts so that technical dilemmas are almost irrelevant. Contrast the technologies of today with those of 100 years ago. Then contrast governments, religion, academics, or economics with those found 500 years ago. Technologies are dramatically different: the social forces are almost identical. Technologies have shown their incredible flexibilities—social forces have merely demonstrated rigidity.

Each region presents a different set of problems. It would be interesting to examine thoroughly the social forces set against an effective implementation of an industrial environmental solution in the various areas of the industrialized world. Unfortunately, such an examination—except a few comments—is beyond the scope of both this book and its author.

There are some obvious problems that are unique to other industrial states. For example, though the Soviet-East European environment has been devastated by their industrial schemes, attempts to change their industrial paradigm are hampered by a political inability to examine its form. Essentially, Soviet industry combines the worst inflexibilities of Adam Smith's mass-production models with Stalinist bureaucracies now complicated by idiotic predatory neocolonial economic advice.

Should the Soviets ever come to grips with the notion that quality production is an environmental issue, and that their environment, while enormous, is clearly finite and must be cared for, and can connect these two issues, they may be first to move to a closed-loop industrial state. The Soviets have some unique advantages: 1) Their industry is sufficiently primitive so they can learn from others' mistakes. 2) They have very strong cultural notions of the common good and this reduces social direction to defining what that good is, 3) They are a highly educated people.

The Western Europeans are the furthest down the road toward an industrial environmental solution in every sociological and technological sense. A careful reader of this book will note that most technical examples of what an industrial environmental future would look like are drawn from the European experience.

The main European limitation is that their technical and political boundaries are different. The acid rain that falls on Norway is formed from the pollution of England. It will be interesting to see if the newly emerging European Community is more than a monetary union—if so, they could address continent-wide environmental problems. The fact that a Community social pact is under consideration bodes well for the future.

The Japanese clearly need an industrial environmental future. Any country (with the size of Montana and 110 million people) that has become the productive center of the planet is certain to be an environmental mess. Moreover, the Japanese have achieved their industrial dominance through an export economy—a plan that is beginning to cause trouble with other industrial nations. They have worked very hard. They deserve to spend some of their gains on themselves. It is preposterous that Tokyo, arguably the industrial center of the planet, should be without a decent sewage treatment system.

Should the Japanese decide to change their industrial scheme to achieve a closed-loop system, they can set directions toward an industrial environmental future on the planet. It would be in their interest to do so for several reasons. 1) The current realization of industrialization is in serious trouble no matter how well it is executed. The Japanese may find that they have become king of a disintegrating hill. 2) Their industrial talent is nearing the stage where their work is becoming meaningless. (After all, how many permutations of the motorcycle—35mm camera, video cassette recorder, laser disc player, etc.—can there be?) 3) They are a nonmilitary industrial state in a militarized world. Many of their customers have gone broke buying weapons. It would be in their interest to give the world a non-military alternative future and provide it with the necessary industrial leadership.

Interestingly, even the Japanese predators seem to understand the practicality of keeping the arms race contained. There is an historical precedent for such a move. The Japanese Shoguns discovered that *firearms made being a Shogun a great deal less fun and successfully had them banned.* (Reread that last sentence!) Somewhere, deep down inside, the Japanese know the arms race is reversible. They could provide an alternative future and induce the world to follow them.

Hurdles of Awareness

Information

A SERIOUS IMPEDIMENT to an industrial environmental solution is the fact that information dissemination has become a branch of the entertainment industry. Television is responsible for a generation of partially informed citizens who are facing decisions requiring genuine understanding. History's most versatile and potentially most profound invention for spreading information has instead become a form of mental anesthesia.

Astute critics of television have claimed that current programming is roughly the same perversion of a medium as if the printing press was only to have been used for making comic books. A comprehensive critique of television, either news or entertainment, is another book. Briefly, the problems of badly informed citizens and the blame borne by television fall into two general areas: lack of any industrial perspective and the lack of science coverage.

No subject is treated worse on television than science. Forget for a moment that science, in the main, is totally ignored by television, and forget that television newscasters routinely assume that everyone in their audience is as technologically and scientifically illiterate as they, and remember when television actually tried to cover a scientific event; the flight of Apollo 11. To anyone remotely inclined to science, the memory is painful but the event itself was so wonderful that the story bears repeating.

The sheer audacity of sending men to the moon was so stunning that the people covering the event spent most of their time trying to comprehend it. Like most people in such a situation, television news teams focused on the irrelevant and the trivial to cope with their lack of understanding. They treated it as a celebrity event, interviewing such towering minds as Spiro Agnew who probably understood even less about the exploration of space than did the television newscasters. They focused on the astronauts, their wives, and families. Mostly it was the picture of the television personality, desperately thinking of something to say beyond some variation of "Oh, WOW!" while trying to fill the airtime between the 150 seconds of blast-off footage and the pictures of astronauts climbing out onto the surface of the moon that will remain the dominant memory of those who watched.

What an utter waste! That airtime could have been used to teach an entire nation the most fascinating aspects of science and large-scale projects. Teamwork could have been demonstrated in a way that would have embarrassed the morality play known as football. Television could have done a piece on how the general method for getting the mission accomplished was decided upon: Who were the persons lobbying for each method? How was the final decision reached? Were personalities involved or was it a purely scientific decision? Did the defeated camps eventually support the final decision? Television could have focused on the huge array of pieces to be made and problems to be solved once the decision was reached as to how to proceed: Were new fabrication methods employed? What materials were used? What were the computer needs? Who were the people who solved the problems and what methods did they use? What unforeseen problems set back progress and why should they have been foreseen? There were thousands of interesting stories but television could only see the daredevils sitting on the top of a tower of explosives. The astronauts were interesting but to focus on them to the exclusion of the others on the team was the equivalent of covering a hockey game and only watching the goalkeepers.

The lack of scientific awareness by persons in television is but the most obvious manifestation of a lack of an industrial perspective. The most common excuse for this deficiency is the belief that the industrial states are passing into the post-industrial-information age and an understanding of industrial problems is unnecessary. That television should treat such nonsense seriously is to be expected. The idea of the information age appeals to the untrammeled narcissism of the typical television personality.

Television people have ample evidence to cite in support of their beliefs: The information age was proclaimed by respected academics (who, of course, had their own self-serving reasons to espouse the glories of information). People in communications were doing better financially than people making steel. Newspapers and television stations showed greater profit margins and growth rates than manufacturing or agriculture. The concept of the information age allowed these people to justify their career choice while basking in the knowledge that they were on the leading edge of the latest trend.

As might be expected, information about any enterprise except the dissemination of information is atrocious. No matter the issue, from business to politics to international events, the pattern is the same. News coverage is as superficial as possible and when the event proves to be too complicated to treat superficially, the reporters turn on each other and have "serious" public discussions about the quality of the coverage. It matters not whether the subject is a political campaign or an airline hijacking, eventually the subject will not be the campaign or the hijacking but the reporters.

Business has more or less given up hope for any sort of valid coverage of their activities. The only time they make the news is when a plant blows up, they are sued, a product fails, or a corporate officer is indicted for some sort of crime. Because business has money, they at least have the alternative of buying time or space to air their views but these are also superficial and likely to reflect a predator rather than a producer perspective. Like science, industrial technologies are seemingly beyond the grasp of news reporters. It is ironic that news about an industrial society should be gathered and spread by technological illiterates but such is the rule rather than the exception.

An industrial environmental solution is virtually impossible unless people are made aware of the connection between technological issues and public policy. This understanding is long overdue. The time for making sound public policy decisions is growing short. The poisons that will affect the water supply are already in the ground. No medium can get the issues out before the public in a meaningful manner faster and more quickly than television.

The greatest service a scientific television network could perform is that of a "Consumer Reports" for infrastructure. Part consumerism, part public affairs, such programming would allow citizens to make informed choices about their future. Programs about waste disposal methods around the world could examine effectiveness, problems, and costs of various systems. Does the mass garbage system of Paris work better than the one in Stockholm? Does either emit hazardous gasses? What are the ash problems? How is the energy used? What were the construction problems? Is either better at solving the odor or noise problems?

The subjects from soil erosion to acid rain, from toxic waste to energy conservation, are virtually endless. The range of solutions is wide but most do not even know what they are, where they have been tried, and how well were the problems solved. Only television can spread such vital information quickly and broadly enough to get informed public action. Television must become more than an electronic comic book!

EDUCATION

MAKING TELEVISION a medium for useful information, promising as that possibility may be, is not going to raise the overall general abilities necessary to implement an industrial-environmental solution by itself. Television is a useful means of spreading information mostly to those who have a background and a curiosity for a subject already.

BACK TO BASICS

ALMOST EVERY EDUCATION SCHEME calls for some form of a return to the basics. What this means is that someone is calling for a more formalized method of preserving archaic tendencies; however, from an industrial point of view, there are some basics that are not taught that severely limit the ability of a person to understand.

PART ONE A course would require that a student learn to build a fire. Knowing how to build a fire means a student must know that fuel, oxygen and ignition are necessary. Not knowing how to build a fire literally means that the student, no matter what else he or she may know, is more ignorant of a basic skill than Stone-age man. Not knowing fire means that most industrial and environmental issues will remain confusing.

PART TWO An examination of agriculture would be required of all students. Because agriculture is the basis of all civilization, ignorance of agricultural issues means that civilization is, at best, confusing. Anyone who thinks milk comes from a supermarket is charitably, an ignoramus. For most of history, education did not need to concern itself with agriculture. Everyone at school was getting the agricultural part of education at home—or at least from a very close relative. With less than two percent of the United States population still involved directly with agriculture, this is no longer true. A simple, but very effective way to teach agriculture would be to trace butter back to the cow. Every step of the distribution and production process should be covered except for packaging machinery which would be covered in part three.

PART THREE A course in basic industrialization would trace a very simple product—such as a plastic-handled screwdriver—all the way back to the mines and oil wells. Each step should lead the student to a greater appreciation of the incredible legacy each simple product is heir to. If a student understands fire, agriculture, and production, everything else will make a great deal more sense.

VALUE HURDLES

IN ORDER for there to be an industrial environmental solution, there must be a reduction in the influence of the most negative effects of predation. The main need is for the predators to get off the necks of the producers.

CAPITULATION TO THE POLICIES OF PREDATION

AT THE ROOT of all predatory activities is the desire to get something for nothing or as close to nothing as possible. Some practices of predation are essentially harmless. Many battles over ownership, for example, have almost irrelevant outcomes except for those directly involved in the battle. Who owns an enterprise is of little importance compared to how the enterprise is run. If the *real* capital (roads, schools, factories, etc.) disappeared tomorrow, the world would be thrust back into the middle ages. If the *money* capital disappeared tomorrow, while there would be furious arguments over who owned things, virtually nothing would change.

In a micro sense, predation causes industrial problems such as high profit margins cutting into research and development or maintenance. Do this long enough and any industrial enterprise will be ruined. Because technology is cumulative it is highly interdependent. The macro effects of predation within single enterprises is the general lowering of industrial potential.

Every time a small element of the productive foundation of an industrial nation is damaged by predatory action, the whole nation is weakened. Persons who use their power to plunder a business rather than build and strengthen it are more than just greedy; they are industrial saboteurs. A terrorist who blows up a power station and the corporate raider who seizes a firm through a maneuver in the stock market and pays the bills by closing the firm and selling the assets have a very great deal in common.

In fact, the bomb-toting terrorist is probably much less dangerous because an explosion in a power station does not destroy potential. Japan and Germany were reduced to rubble at the end of World War II but were able to rebuild because the industrial potential of the people had not been destroyed. Barring nuclear destruction, anything that has ever been built can probably be built again—especially if the people are still around who built it in the first place. When a factory is closed, the people are dispersed and retrained, and research and engineering staffs are broken up; a living, productive organism has been so thoroughly destroyed that only one attempt in 100 will succeed in replacing it.

The predatory damage of irresponsible business practice is merely a warm-up act for the real social damage caused by the primal predation of militarism. Predation at the business level leads to economic and industrial decline. Predation extended beyond industrial decline leads to militarism. Militarism not only leads to war but causes economic upheaval and destruction of creative institutions.

THE PRICE OF LIBERTY IS ETERNAL VIGILANCE

MOST FORMS of predatory activity are destructive but none is as destructive as militarism. While warfare is clearly a threat, the mere presence of military institutions in a society constitutes a hazard to that society's well being. The military is truly a profession apart. Where most of a society's citizens believe that killing other humans is a serious crime, the job of the military is to organize people to be ruthless and efficient killers. More than that, the military must make people believe that killing under conditions defined by the military is a noble and honorable activity.

The armies of democracies are supposed to be hirelings of the state willing to kill and die for whatever objective that state deems necessary; but armies subject to the wishes of the state are a rather new invention. For most of history and in many countries today, the army is the state. Fighting was a noble activity because only noblemen did it. Warfare was the activity of the ruling class.

Deciding what was worth protecting and who should die was a right the aristocracy reserved unto itself. The absurdity of allowing the foreign service to decide who should be the enemy is merely an example of an antidemocratic, aristocratic tendency. Making soldiering a profession did little to change that. In organization and function, armies are not democratic.

The mindset of a military officer is that of a despot by training. As a result, professional military organizations and democracies are not natural allies and coexist very uneasily. Not only do military minds exhibit a tendency toward totalitarianism, they also tend toward the arrogance of their aristocratic roots. They often believe that they can operate a country as well as their military units and have, more than once, succumbed to the temptation to try. As a result, true civilian rule of countries is very rare.

GUARDING AGAINST PREDATION

IT WOULD BE FATUOUS to assume that this book would have a good solution for the problems of excessive predation. The great teachers from Jesus of Nazareth to Thorstein Veblen have addressed the problem without much success and anyone who believes the problem will disappear any time in the future is probably a hopeless optimist. Although this may be an example of utter futility, the problems of excessive predation must be discussed because they are so serious. The nuclear age has rendered predatory solutions to economic problems impossible; nations can no longer go to war to solve problems at home.

The bright spot in the gloom of predation consists of the overwhelming realization that Germany and Japan are not militarized—most of their populations oppose war. The cheerleaders for war are found in the industrial states with major economic problems such as the England the United States. The opponents of military solution are the new industrial superpowers. Even, or especially, the predators slavishly imitate the rich. If the very rich countries of Japan and Germany have come to the conclusion that predation is the road to ruin, maybe the rest will follow. One can hope!

MILITARISM

PUTTING A STOP TO MILITARISM ITSELF will be the harder still. Because of the existence of predators, the world is a dangerous place and a certain level of defense is necessary. Moreover, militarism has at its disposal some powerful symbols which make rational discussion about what is an appropriate level of defense nearly impossible. Wave the flag, play the martial music, and hold solemn memorial services for fallen warriors and almost any level of militarism can be justified.

Add to symbology the very real fear that without the economic stimulus of military spending, a depression would occur and rational discussion is impossible. *An army that does not feed on an enemy must feed on its supporters.* Excessive militarism is a recipe for economic decline.

Patriotism is the value system of the predators, and rooted as it is in the notion of defending one's turf, it is a very powerful and natural system indeed. But producers also have a powerfully rooted value system that is at least as valid as any patriotic instinct and provides an excellent value alternative. Veblen defines the alternative producer values as the *parental instinct, the instinct of workmanship, and the instinct of idle curiosity.*

The *parental instinct* is summed up in the desire to make the world a better place for one's children. This may not be a universal characteristic but it is very high on most value scales. Industrial environmentalism provides an outlet for this desire while giving an industrial alternative to militarism.

The *instinct of workmanship* is amply demonstrated by the pleasure everyone takes in a job well done. No one likes to build a lousy building if there is a chance to build excellently. The elegant solutions necessary for industrial environmentalism would provide a perfect outlet.

Of the three, the *instinct of idle curiosity* might seem the most frivolous but it may be the most important. Many elements of the industrial environmental solution have not, unfortunately, been invented. Idle curiosity, the desire to ask "what if," is an anti militaristic sentiment. Militarism does not brook idle questions: it demands unthinking obedience. As it stands, the instinct of idle curiosity is perverted by militarism and makes the world an ever more dangerous place.

REDEPLOYING THE MILITARY

NO MATTER HOW MUCH GOOD it would do, eliminating or significantly reducing the military is not possible. Arguing that militarism damages civil liberties, causes governments to lie extensively, and generally diminishes the freedoms of citizens has never done any good in the past.

The militarists argue that to defend freedom, a country must sacrifice those very freedoms. In spite of the logical absurdity of this argument, people have bought it for a long time. Arguing that militarism is waste will do no good. Pointing out that the only thing worth fighting over is probably the Middle East oil fields will accomplish little. Waste has always been a status symbol of wealth and power for the predators who never view it as an evil. Even arguing that the most dangerous military is the one closest to the discussion is likely to fall on deaf ears though both logic and history bear it out. People side with the home team no matter how futile or rotten the organization, as the existence of Chicago Cub fans demonstrates.

Merely pointing out the flaws of militarism is doomed to failure because of the institutional realities. Cutting 90 percent from the military budget, possible from a strategic standpoint, is not only politically unlikely, it would, in the United States, destroy whatever is left of the industrial economy. Besides, no matter what one happens to think about militarism, throwing that many people out of work seems a bit excessive. Moreover, the military is filled with excellent people, many of whom are producers anyway. The point is not to throw these people out of work, the goal is to redefine their mission.

It would be useful to make environmental protection a military activity. They made many of the worst messes. They are equipped to handle large projects. They already have extensive organizational abilities. They are able to motivate young people to do hazardous work. They know how to shove large budgets through legislatures. They have extensive ties with the manufacturing sectors of the economy. Most importantly, they are an approved agency of collective action.

To some extent, this has already happened. The Coast Guard has been assigned to handle offshore oil spills. The trick would be to expand this role 100-fold.

The advantages would be stupendous. As it stands now, the military must perform constant exercises. No matter how much fun war games may be, eventually they get old. Keeping a group of young people at a warlike pitch is really not possible. By assigning environmental protection as the primary role, the military would be constantly evaluated as to their abilities to accomplish a mission without going to war. This would keep them sharp in case their warlike abilities were ever needed.

The advantages to those who would be recruits are even more profound. If the role of protecting the country was expanded beyond the current definition of killing a foreigner whom your government has decided is the enemy to include the active preservation of the environment, the resulting citizen would be a model to behold. A person who has dug out toxic waste barrels by hand so as not to cause further damage will be careful never to make such a future task necessary. A person who has helped build a country's infrastructure will supports its future maintenance.

Many tasks necessary for environmental repair can best be done by hand. For instance, by the time acid rain has been brought under control, the forests covering whole parts of continents will have been damaged. It will be necessary to prune salvageable trees, cut dead trees, and replant. This can only be done properly by a trained army using hand methods. Imagine the good that could be accomplished if environmental protection was a part of the indoctrination of boot camp. It must be taught that environmental protection is as much a matter of life and death as war.

The environmental damage may be so severe that serious consideration should be given to universal conscription—make the stint in the environmental army a rite of passage to citizenship. If everyone had to serve two years between 18 and 20 without exception, regardless of race, sex, or natural ability, huge projects are possible. Universal conscription would probably cost less than the current professional army, even if every conscript was promised as much education as he or she could handle upon completion as a reward for service.

The current organizational military hierarchy, which is currently devoted to procuring ever more technologically sophisticated and bizarre weapons systems could put their talents to work making sure that environmental systems are designed properly. Again, this would have the advantage of making them more organizationally sharp. Buying weapons that cannot be used means there are no meaningful tests of performance. Success in protecting the environment can be measured.

If the military were to devote 90 percent of its resources to environmental protection, it probably will be protected.

Elegant technology will solve development problems virtually anywhere. The country with the most elegant technology will be the country with the most international influence. Because of the widespread poliferation of arms, the only war that can ever be won is a war of development.

It would be interesting to hear the President of the United States include the following in a speech some day—and mean it:

"Our nation has decided that any further escalation of the arms race is absurd. Warfare only destroys our own economy. We are already so dangerous that we cannot further comprehend it. We have discovered that we are losing a much more serious war—the war we accidentally declared on our own systems of survival. As of this day, I am redirecting our forces to this new front. It will be a long battle. Our own ignorance has caused much damage and we have been doing it for a long time.

"This is not to say that we have become suddenly weak. We still can destroy the planet with the touch of a button. Sad to say, this problem will never go away. But we will be stronger. We will invent and build the solutions to the mistakes of our past and thereby demonstrate conclusively the superiority of our economic systems. We will prevail. We call on others to join us in these efforts. Their environmental damages will eventually affect us as well. Even if they do not join us, we will proceed. We are strong in armed might. We will be strong in sophisticated and elegant technology."

This is the minimum commitment in the industrial struggle.

POLITICAL HURDLES
THE NEED TO INVENT A PRODUCER POLITICS

INDUSTRIAL ENVIRONMENTALISM may be a plan that does not involve much political change to implement, but it does involve change. In the end, any collective action involves politics and the industrial environmental solution is a collective action.

Predators are fascinated with solutions that involve litigation. Unfortunately, lawsuits solve almost nothing except the income problems of lawyers. This is an obvious waste of time since when it comes to toxic waste, everyone is to blame. Worse, while the suits drag through the courts as the legal system struggles in vain to assess blame, the toxins spread further into the groundwater.

Legislators think they can delegate their responsibility for effective solutions merely by appropriating money. They fund agencies staffed by professionals who evade their responsibility for action by claiming intellectual caution when in fact the real problem is that they do not want to hamper their management careers by getting their hands dirty solving a problem. Public servants have become anything but.

Increasingly large numbers of civil servants have been allowed to accomplish less and less while being paid more and more. Everyone seems to think that they can hire someone else to solve the problem. This daisy chain gets larger and larger until all the funds have been spent. What is even more ridiculous, this irresponsible buck-passing is counted as economic activity by the fellow government narcissists in the agencies responsible for those economic records.

Only a hopeless optimist could assume that real collective action through such institutional government obstacles is possible. The industrial environmental solution requires industrially aware legislation, real environmental action, and a whole society-full list of new products and improvements. Since change is necessary for future survival, the institutional obstacles must be eliminated. In a political sense, that means elective action.

The producers' first step should be a demand that elected officials have some grasp of the realities of the twentieth century; its science and industry. Just once it would have been fun to hear Bush and Dukakis answer a scientific question such as "Describe your understanding of the second law of thermodynamics and how that understanding effects your position on energy planning, energy lending, and foreign policy." It should not be too much to expect a President in the twentieth century to know the second law of thermodynamics. A President, or any national elected public figure, should also be expected to understand geometric growth, the basic environmental sciences, international resource distribution, and indicators of industrial competitiveness. In fact, not knowing these things should pretty much disqualify any candidate.

Serious consideration should be given to electing a leader with real producer credentials. The problems facing the planet are simply too serious for another actor or lawyer. A great scientist would be tempting but scientists rarely understand the problems of production or the organization of large efforts. A better place to look might be the world of advanced-technology manufacturing. If the company is a multinational enterprise, it would be well if the candidate had served some time overseas and learned the language. It would be useful if the candidates had an interest in history, real scientific, inventive, or industrial credentials, and enough knowledge of people to be an effective manager.

There is just no escaping the fact that Industrial environmentalism involves industrial planning. Failing to elect a leader who understands the process does not obviate the need for such planning.

What is important is the form industrial planning takes. A great deal of idle talk has already clouded the issues and aims of industrial planning. One of the worst proposals is that a political commission be set up to decide what is a sunrise or growth industry for the future and favor that industry with development funds. Probably no group in any society is less qualified to make such decisions than politicians. Not only do politicians generally lack the education and other prerequisites to understand the nature of industrial issues, they are usually anti-industrial by temperament.

To expect such a group to choose between sunrise and sunset industries is not only futile, but there already exists a much more useful mechanism for such decisions—the market. The market may have problems with accurately setting prices, determining the value of money, and so on, but it lacks peer when it comes to determining what consumers believe to be desirable.

In point of fact, because industrial planning is to concern itself with the issues propping up the market economy, it could be viewed as a market economy subsidy. This is hardly a new phenomenon, market economies have always been subsidized—most prominently by the environment. It has not been common to think about the environmental contribution to the economy because as little as 100 years ago, almost no one could even conceive that humans had the power to destroy nature. Even 40 years after such destruction became literally possible, most people are still unable to comprehend it. At least some are going to have to think about this subsidy and make plans for its continuance. This is the role for industrial planning.

It is also an apt role. If the economy of the biosphere were to be considered as the operation of a major company, the market economies would be the equivalent to the research and development departments. Research and development in a company is often subsidized heavily. Industrial planning should concern itself with the question of what is to be done. The market economies can decide best how it should be done.

Economic Hurdles
Is Industrial Environmentalism Affordable?

THE MOST PRACTICAL OBJECTION to industrial environmentalism is, of course, its cost. There is no way of accurately measuring the cost of a project that could take 50 years to complete, but be assured, it would be very high.

Placing a monetary figure on any long range plan is usually futile. But at the risk of sounding frivolous, it should be recalled that *this is only money.* The only valid question concerning cost is whether a project is environmentally affordable—the very question that has not been asked for most of human history. In almost every respect, the sheer size of this project is not a fatal problem, it is its greatest asset. It must be remembered that the goal of industrial environmentalism is not merely to correct the industrial mistakes of the past, it is to be a form of economic stimulation.

Industrial environmentalism is a huge project. If a society is attempting to clean up a toxic waste problem, it simply cannot afford to suspend the efforts to wait out the economic problems caused by the latest recession. Playing the boom-bust cycle on Wall Street may be great fun, lead to the accumulation of vast holdings, and even have the benefit of shaking the industrial losers out of the system but these games cannot be allowed to intrude on long-range industrial environmental projects.

The solution may be simply to make public investment without triggering the issue of debt. As the current Federal Reserve System works, increases in the money supply lead to debt formation. The money supply must increase to compensate for the growth of economic activity. Logically, the projects associated with the industrial environmental solution could and should be separated from the normal market mechanisms.

Industrial environmentalism is industrial planning. As there are no markets between the present and the future, such planning is, by definition, a *supra-market* endeavor. If it is in the best interests of the planet to proceed with supra-market projects and it is also necessary to increase the money supply, the logical answer is to combine these activities.

Instead of passing out new money through the banks as debt, new money could be paid *directly* through the industrial environmental projects. They could serve as the *starting point* for the increased money supply as easily as banks and the benefits would be enormous. If debt was not created, there would be no reason ever to panic. Because new money would start at the point of economic pie expansion, increases in the money supply would be far less likely to become inflationary. Projects would not have to wait funding but would only have to prove that the necessary resources, human or otherwise, were available. *If there was a problem and a solution and a plan, money would be created to get the project started.*

Market economies are absolutely superb in catering to the wants of humans. They fail when it comes to meeting real social needs. The answer is not to replace a market economy with a planned economy—socialist or otherwise, or vice versa, the answer is to have *both.* There is clearly enough talent and need for both. If a society has a planned segment that is responsible for environmental human needs, the market economy will flourish on its own. Unfortunately, most people in power seem to think that market and planned economies are an either/or rather than a *both* proposition.

The realization that society needs both a vigorous public and private economy would solve the major social problem of what to do with mature products and production. An obvious example would be farm tractors in the United States As of 1986, there were no domestic tractor producers showing a profit. Overcapacity plagued the industry. Some would argue that the industry should be allowed to die a natural death. In reality, the ability of a nation to make farm implement amounts to its ability to eat. Some things are necessary even if they are not profitable. The industry has passed from growth to maintenance in its life cycle. Innovation needs are replaced with stability—the builder with bureaucrats.

At this stage of industrial existence, there is almost nothing worth gaining from competition. Competition that does not lead to product improvements leads directly to waste. Trying to extract profits from necessary but static industries is a prescription for destruction. A very strong argument can be made for converting such industries into public, nonprofit companies.

Let the markets decide what new industrial endeavor should succeed but let the public decide *which of these aging monuments to past creative genius should be preserved for the simple reason that everyone has grown to rely on their presence.* Such conversions to public ownership should be thought of as preservation. Political slogans such as nationalization only confuse the issue as to what of the industrial landscape must be kept.

Even if the industrial environmental solution could be funded without incurring debt there is still a role for the social aspects of taxation to pay for these projects. Not only would it lend an air of fiscal responsibility that may be necessary to stave off total decadence, it could focus attention on the real costs of abusing the environment.

Because energy is the central element of industrialization, it could have an even greater role in the financing of an industrial environmental solution. Should created money outstrip real industrial gains to the point where confidence in the currency begins to waver and inflation begins to break out, confidence could be restored by valuing money in the most important conversion of all—energy. Money could be valued in BTU equivalents. Such money would be astonishingly accurate. Building windmill farms to harness that form of solar energy would not only make that nation richer in absolute terms, those riches would show up in an increased money supply. No more powerful incentive for solar power could be created than to make installations literally a justification for printing money.

Bold strokes such as changing the way the supply of money is expanded or even not-so-bold changes as increased energy taxes will require a sea-change in political will. Even if such changes do not take place, it will still be possible to get a very good start on an industrial environmental solution merely by redeploying assets. The juiciest target for redeployment, of course, is the military. Not only does the military get more than its share of the money, this huge supply of money enables the industries associated with military production to hire the cream of the industrial talent. This is the talent that is critical to the necessary industrial environmental solutions.

By simply giving the military a more important mission than provoking confrontation, important elements of the industrial environmental solution could be started with no changes to the world's financial systems whatsoever. The only real question is whether humans are more willing to change their minds about militarism or money. Both are deeply entrenched in human institutions so change to either will be very difficult. The good news is that changes in either one will get things started. Changing the thinking about militarism may be preferable because of the positive social spin-offs such as reduced world tensions combined with the more enlightened social, economic, political, and educational institutions necessary for sound environmental and industrial planning. Changes in both economic and military thinking would be best of all and eventually mandatory. Mandatory or not, effecting these changes will be a Herculean task.

CHAPTER FOURTEEN

THE POPULIST PERSPECTIVE

IN THIS BOOK, I have tried to be as fair and evenhanded as possible and I believe I have succeeded. My point of view has a name. It is populist—Midwestern, prairie, Donnelly, Veblen, populist.

Confessing to being a populist is a strange admission. Whatever the dictionary definition of Populism might be, the common-usage definition of a populist in America is that of a low-brow political opportunist who appeals to the basic instincts in humanity to get votes.

Using that definition, persons as foul and diverse as former Alabama Governor George Wallace, David Duke, and Argentine strongman Juan Peron have been mislabeled "populists" in the American press. One would not casually link one's political philosophy to any of those persons, nor even to people such as William Jennings Bryan and Huey Long—both of whom were much better examples of Populists.

I grew up on the plains of Minnesota and North Dakota and we are much more proud of our populist history—and with good reason. The writer-historian Ignatius Donnelly, who became the national Populist Party's "Jefferson" by writing the fiery platform introduction of 1892, was Minnesota's first Lieutenant Governor. (Good Populists might argue that Jefferson was the Populists' "Jefferson.") Both Donnelly and his utopian community called Ninniger are honored as place names in St. Paul.

North Dakota's most successful outbreak of populism was known as the Non-Partisan League. It took over the control of the state house and instituted arguably the most progressive government this country has seen. In Minnesota, the populist party was called the Farmer-Labor party. It ran the state during most of the Great Depression and merged with the Democrats following World War II to form the Democratic Farmer-Labor (DFL) Party, by which name it is still known.

As a result, people here in the heart of American are understandably miffed by the appropriation of the populist label to describe despots and hicks. Populism is probably democracy at its finest because it combines political and economic democracy—a rare combination. As for populism's hick image, exposure to the facts indicates that populism is a sophisticated, highly intellectual body of thought.

THE CULTURAL MANIFESTATIONS OF POPULISM

DESCRIBING CULTURAL POPULISM accurately would be the subject of another book. Actually, the fans of Garrison Keillor know those books have already been written. Even growing up in the same area of the country that helped spawn the first great outbreak of populism is not much help. Closeness to a subject brings the problems of perspective. With that disclaimer, I shall attempt to describe populism's main cultural features.

The unifying characteristic of cultural populism is a belief in common sense. Scholars like to call this philosophy of common sense "pragmatic rationalism." Whatever it is called, it was born, in America, of necessity.

The question could be asked, "Why should people not stand up in a sporting arena?" There are many answers to such a question but the preferred populist response would be, "One does not stand up in a sports arena because if one person does it, soon everyone must stand. When that happens everyone will still have the same relative view of the proceedings as when they were seated—only now they will be less comfortable."

It is simply astounding how many profound social issues can be argued using a similar reasoning. There may be other better forms of reasoning, but the appeal to common sense has the advantage of convincing a wider assortment of persons. If, for example, a devout Roman Catholic were told that he should not stand up in a sports arena because the Holy Father had prohibited such an action, the issue would be settled for that devout Catholic. For the rest of humanity, however, the issue would not be settled at all. In fact, many would stand up merely to prove that they were not Catholic.

People who live outside the United States often forget what a gigantic problem cultural diversity can be. There are several reasons for this—the best one being that cultural diversity has many advantages. There is absolutely nothing that is typically American. It does not take a race riot to prove the point either. Out on the high prairie one can find tiny hamlets with a Polish Catholic Church on one side and a German Catholic Church on the other. Because members of both would be classed as white, European, and Catholic, census and other official data would ignore the distinction. But this seemingly tiny difference is so deeply felt that official church rules about "one parish per town" have been suspended. In America, just getting on with one's neighbors requires major effort. Cooperation is almost impossible.

Labor unions are quite rare in the United States. Cultural diversity plays a key role in this fact. People differ profoundly on questions such as: Who is entitled to the rewards of labor? When is work too hard? Who should lead? How are leaders rewarded? What is the worth of physical labor vis-a-vis mental skills? What is the role of privilege, such as nepotism, in a scheme of rewards for merit? If merit is the sole judge, which merits are to be most valued?

If white, European Catholics in small towns cannot figure out a way to worship together, it should come as no surprise that Utah Mormans, Miami Cubans, and New York Jews can find ways to argue about how a society's economic system should be organized.

This explains why pragmatic rationalism was the philosophy made in America. With literally thousands of cultural and religious backgrounds, Americans were forced to come up with a philosophy that was common to all. Only by appeals to common sense is positive collective action possible in America. Americans are at their absolute best when all leave their different ideologies at the door.

If the philosophies of pragmatism nurture the collective good by providing a cultural "Esperanto," it was reinforced by the forces of nature. Populism became strongest where nature was harshest. Only the most pragmatic individuals survived the high prairie long enough to form the populist parties. Populism was a matter of survival. The prevailing sentiment was "It's all right for a good populist to be a utopian dreamer, just be practical about it!!"

Populism, at its best, takes the most attractive idea available and tries it out. If the attempt does not produce the desired results, then something else will be tried. Such a pragmatism, it should be noted, does not produce the moral illiterates that call themselves pragmatists in the latter part of the twentieth century. The pragmatism of populism is directed outwards. Some social notions of pragmatic populism would include: Just social conditions are necessary for social order; Social order is good because it produces prosperity; Prosperity for the greatest number of citizens leads to social order; Justice is necessary for prosperity; and, Ideologies that hinder the prospects of social justice hinder the prospects for prosperity and should be discarded simply because they do not make sense.

Populist pragmatism finesses the question of equality in an interesting way. To say that everyone is equal runs contrary to much evidence. Person simply have different skill levels, abilities, and motivations. People also tend to stratify. Populists object to the arbitrary definition and the artificially small sizes of elites. They particularly object to closed elites. Ignatius Donnelly summed up the stratification problem best when he wrote in *Caesar's Column*, "There is as good stuff in the ranks as ever came out of them."

Moreover, a good populist believes that it should not matter what skills and abilities one might possess. Everyone can contribute to the community. A just society is that society which asks of everyone equally to give their best effort. Performance within occupational types should be the indicator of stratification rather than occupational category. "Who, after all, " the populist might ask, "is more important to your daily life—your garbage collector or your Senator? If the answer is the garbage collector, why would you treat the Senator better?"

Strange as these populist principles may sound to those who seem to believe that it is every man for himself, they are in use across a wide segment of the American landscape. It could be argued that most Americans are cultural populists. It is certainly true that from the Appalachians to the Rockies, cultural populism persists no matter the current status of political populism—which, while currently staging a comeback, is much less common.

Political Populism

THE RISE IN POLITICAL POPULISM coincided with the explosion of intellectual and industrial activity between 1870 and 1940. In those 70 epochal years, humans learned how to make steel, extracted aluminum from bauxite, developed petroleum and organic chemistry, discovered nuclear power, harnessed electricity, learned to fly, perfected mass production techniques, invented radio, television, and the telephone, and built the roads, bridges, and railroads necessary to settle a continent. It is difficult for those of us born after World War II to have any appreciation for the changes wrought on the world by our grandfathers.

To say that the institutions of government, education, religion and banking were unprepared for this intellectual-industrial explosion would be to stretch the definition of understatement. They were unprepared because none participated in these changes. This intellectual-industrial explosion was produced by persons who were mostly uneducated, poor, and from families outside the social elites.

In this day when being technologically illiterate is fashionable, especially among political progressives, it is hard to remember that during the most progressive period of America's history, technological progress and optimism were linked to advanced political thought.

Whether advanced social conditions led to advanced technology, or the other way around, is a classic chicken and egg argument because they happened at largely the same time. Whatever the arrangement, the Industrial Revolution triggered a political response. In England, the Enclosure Act combined with a royalist-feudalist history had pretty much settled the question of who was going to own things. In such a settling, the industrial revolution merely changed the nature of work, not the power arrangements. Such conditions provoked Marxism.

If the Enclosure Act triggered the conditions that produced Marxism, the Homestead Act was the trigger for Populism. America was not England. Land was to be had by anyone willing and able to work it. There was a big catch, however. Fertile as the lands offered by the Homestead Act were, they could not be farmed without the tools provided by the industrial revolution. To be a farmer on the high prairie meant being a multiskilled small businessman.

If Marxism was the political movement of landless peasants, displaced artisans, and the toiling industrial masses, Populism was a movement of landed businessmen who own a means of production. Marxism concentrated on working conditions. Populism sought to make business conditions fair. The rallying cry of a Marxist is that labor should be rewarded. The populist cry is that enterprise should be rewarded.

In the battle for the hearts of Americans, Marxism did not have a chance. It was foreign. It was anti business. When the Populism movement crested in the American Midwest, it was denounced by the Marxists as a bourgeois movement. They were correct: Populism was bourgeois.

Bourgeois or not, Populism was a very progressive movement. America was born in a revolution and had survived a civil war fought over economic and social organization. A huge social land development scheme, launched on a continental scale had combined with the greatest changes in production methods the world would ever see to produce a progressive outbreak.

Like progressives of any epoch, the technological farmer-businessmen of the high prairie were out ahead of the cultural institutions of government, finance, and education. Populism was to be a political movement demanding honesty and rationality in government, relevance in education, and fairness in finance.

The main frustration was economic. When economic activity increases at the explosive rate seen from 1870 to 1940, the money supply must also increase to lubricate the wheels of commerce. Because America was on the gold standard for most of that time, the money supply only increased as fast as gold could be found and mined. This was not nearly fast enough to accommodate an industrial explosion. The result was that growth, inventiveness, and enterprise were constantly choked off. Let me quote Donnelly:

"Gold and Silver are the bases of the world's currency. If they are abundant, all forms of paper money are abundant. If they are scarce, the paper money must shrink in proportion to the shrinkage of its foundation; if not, there comes panics and convulsions, in the effort to make one dollar of gold pay three, six, or ten of paper. *For one hundred and fifty years the production of gold and silver has been steadily shrinking, while the population and business of the world have been rapidly increasing*. (His italics)

"Take a child a few years old; let a blacksmith weld around his waist an iron band. At first it causes him little inconvenience. He plays. As he grows older it becomes tighter; it causes him pain; he scarcely knows what ails him. He still grows. All his internal organs are cramped and displaced. He grows still larger; he has the head, shoulders, and limbs of a man and with waist of a child. He is a monstrosity. He dies. This is the picture of the world to-day, bound in the silly superstition of some prehistoric nation. But this is not all. Every decrease in the quantity, actual or relative, of gold and silver increase the purchasing power of the dollars made out of them; and the dollar becomes the equivalent for a larger amount of the labor of man and his productions. This makes the rich man richer and the poor man poorer. The iron band is displacing the organs of life. As the dollar rises in value, man sinks. Hence the decrease in wages; the increase in the power of wealth; the luxury of the few; the misery of the many."

The central Populist economic argument was that when money is in short supply to begin with, and then is concentrated in a few hands, the result is a social, political, and economic catastrophe.

The Populist program called for increased money supplies coupled with restrictions on the methods for accumulation. Some populists wanted to increase the money in circulation by including silver as legal tender. These were called "bi-metalists." The official Populist position (including Donnelly above) called for paper money.

Merely providing the economy with more money was not enough. Populists wanted persons to a decent living without having to spend their whole lives attempting to defraud their neighbors. More than that, Populists wanted protection from those who did. They demanded government regulation of trusts, railroads, and food standards.

Populists had varied success at the ballot box. They never won a presidency but won several Senate seats, some house elections, and succeeded in taking over several state governments. Make no mistake, when the Populist movement was in its heyday, it was a radical movement. Because it grew in areas that have best been called the "Internal Empire," Populism, in many ways resembled an anti colonial uprising. The battle between Wall Street and Main Street was much more than a figure of speech.

Eventually, of course, the proposals of the Populists became no more radical than the light bulb. From paper money and discounted home mortgages to government regulations on large business, the populist agenda was largely adopted. Ironically the incorporation of populist notions combined with a variety of other factors to kill populism as a separate movement.

Many of the most sacred notions of the Populist agenda, such as abandoning the gold standard, were adopted during the administration of Franklin D. Roosevelt. Populist victories were relabeled "liberal" victories. This was not a sinister appropriation of ideas. Populist ideas became liberal ideas because populism was losing its base. People were moving from the farm which effectively shrunk the natural base of the movement. More importantly, the leading lights of populism seemed determined to embarrass themselves in public. Henry Ford had his anti-Semitism and union-busting. William Jennings Bryan will be remembered best for his prosecution of John Scopes for teaching the theory of evolution.

In Bryan's defense, one must recall that Darwinism had a social side that was used by the rich to justify the misery of the poor. Social Darwinism justified a hideous social order by claiming it was a natural phenomenon. In Bryan's mind, anything used to such evil ends had to be evil. That Darwinism contradicted the literal interpretation of the Bible may have not been Bryan's biggest dispute with the teaching, but with a backing of millions of Biblical literalists, Bryan chose to attack Scopes for heresy, and in so doing, doomed Populism as a movement of, by, and for people of great genius. From Bryan on, populism became defined as any political movement of hicks. Interestingly, Bryan was done in by Darrow, a Socialist. Bryan, the great man of peace, was discredited by an attack from the left.

In the 1990s, populism is making a comeback. Now it is the liberals who have an image problem. Political progressives need a home. More importantly, it is the attack on the populist agenda that is at the root of the economic problems facing the American middle class. The middle is groping for answers and increasingly they find them in their populist roots because populism was always about middle-class values.

The rational pragmatism of populist philosophy applies to many of the political and economic arguments raging today. Deregulation of essential services such as the telephone has proved disastrous. Deregulation of air travel has frightened thousands of flyers. High interest rates have priced thousands out of a decent standard of living—especially when applied to housing. High interest rates on industry have devastated productive enterprise throughout the globe. Populist victories produced the American middle class. Populist defeats have been an assault on that same middle class. Not surprisingly, the middle is shrinking.

Americans think that the middle class is ordained of the gods. They forget that it was a product of their grandparents, and their wisdom. Many societies exist throughout the world with no real middle class. Like anything else, the middle can be destroyed.

In America, the middle is under siege: the political middle, the economic middle, and even the geographic middle. Charts of American prosperity in the 1980s showed that the 34 states in the middle of the country were in recession during a "growing economy." Great societies are produced by the middle. An assault on the middle is an assault on the social fabric.

The attack on the middle comes from an odd grab-bag of political bedfellows. From the political right the assault is a combination of Wall Street dealmakers who busy themselves destroying the futures of small midwestern cities, and their neofeudalist apologists who concocted the utter barbarism known as monetarism. From the political left, the antitechnology, antibusiness, postindustrialists propose rationales for why this destruction does not matter.

The return of populism in the 1990s may be nothing more than liberals who are seeking a semantic way to shed their discredited label. These may be small points but there are some significant differences between populists and liberals. Liberals are anti-enterprise, anti-technology, and discount the importance of agriculture. Populists value business conditions favorable to prosperity, are comfortable with technology, and believe that agriculture is the foundation of civilization. Liberals fear the excesses of religion. Populists are often quite devout. Liberals focus on process. Populists focus on product.

So, issue by issue, event by event, the New Populism gains strength. The destruction of American agriculture and basic industry, plant closures, bank failures, the trade deficit, eight percent—or worse—home mortgages, insider trading and other Wall Street abuse, deregulation, government corruption, the diversion of tax revenue to debt service rather than human service: these, and more, provide ammunition for the new populists.

WILL NEO-POPULISM TRIUMPH?

IT IS INTERESTING to speculate about the future of populism. After many years of backwardness and sheer ignorance in the public arena, America is long overdue for a progressive political movement. History usually swings between political extremes and by that theory, history favors a progressive outbreak. American conservatives have just about torn off their arms holding back the hands of time.

There is nothing inevitable about a return to consciousness in public behavior, however. Progressive political outbreaks have a very sporadic history. There are essentially two reasons for this: It is easier to keep things as they are than it is to change things; and, it is easier to destroy than to build; but, ignorance and backwardness cause misery and eventually people will risk building the new social forms necessary to eliminate that misery.

Populism is not the sole claimant for a progressive return by any stretch of the imagination. Old-fashioned American liberalism could stage a comeback if it could somehow shake off its preoccupation with non-issues.

In America, populism may have the inside track for it has many interesting, attractive, and relevant features that address themselves to the most pressing of modern dilemmas. They are:

•Populism has historically been the centrist political position. The center contains the most people and, in a democracy, that occasionally means power.

•Populism is American. No matter what theoretical value socialism may have ever had, it is foreign. Even at the height of the Great Depression, less than one million people voted for the Communist Party. Americans will never be of the mind that Marxism, Socialism, or Communism is progressive. They never have believed it was progressive and 70 years of anti-Marxist propaganda have only ingrained such a position. Even the Marxists no longer believe that Marxism will solve anything. Besides, if past performance is any indication, socialism was better at promising a progressive future than actually delivering on one. Even the progressive newcomers, such as West Germany's Green Party, will not survive a transplant. Populism is as American as the "Tin Lizzy," Mark Twain, and the McGuffey reader. It has won in the past because it is home-grown, which means it is particularly relevant.

•Populism is pragmatic. Because populists are on the lookout for any good idea to try—no matter its source, it is in the position to adopt any progressive suggestion.

•Populism is appropriate to the circumstances. Because current economic problems are almost identical with those which caused the first outbreak of Populism in the 1890s, a second outbreak is more possible.

The signs are all around us. Politicians are taking to calling themselves populists again, but a new populism will never be identical with the original. While the problems that currently face humanity, and especially the United States, are clearly analogous to those that gave us the first incarnation of populism, they are different. Useful lessons can be drawn from the past, but they should not be overdrawn. Moreover, Populism was far from perfect. Like almost anything else, it can be improved upon.

SUPPLEMENTAL POPULISM READING LIST

IF MY BOOK incites a further interest in Populism, the following list of materials should give the reader a good background in the history, ideals, goals, and monetary theories of one of America's most positive movements. Of course, Lawrence Goodwyn's book mentioned in chapter six is the logical starting point.

Ole Rolvaag, *Giants in the Earth.* Yes, this is a work of fiction, but I will vouch for its realism. More than 20 years ago, I survived two winters in western North Dakota in a modern dwelling with central heating. The experience was traumatic enough so that even today, I shudder merely thinking about it. Rolvaag paints a magnificent picture of the people who were brave enough to go to such a place with what they could carry in a wagon—and survive, and sometimes prosper. These people may have been called "hayseeds" but they were truly giants in perseverance, courage, and intellect. It is the story of the kind of people who would go on to form the populist parties.

Robert Morlan, *Political Prairie Fire.* An excellent history of the rise of the Non-Partisan League in 1910s North Dakota. It was written before all the participants had died and is really a source book for all other histories of the League. Best of all, it was the source of a good movie called...

Northern Lights. Winner of the Golden Palm at the Cannes Film Festival, *Northern Lights* excellently conveys the conditions that spawned such a progressive movement as the Non-Partisan League out in the sparse, barren lands of the high prairie. Worth ten books.

Ignatius Donnelly, *Caesar's Column.* This is real source material. Donnelly was one of the self-taught scholars of the populist movement. Sent to congress from Minnesota, Donnelly apparently spent most of his time reading in the Library of Congress. It is no wonder he was considered such a radical. The idea that leaders should be well read died with Jefferson. It has been seen only sporadically ever since.

Caesar's Column is a Utopian-Dystopian romance novel that became a 250 thousand copy best seller. It was set about 100 years into the future which was 1988. It is interesting to see how many predictions have come true. Combining dire warnings, a Utopian vision, and an action-packed love story, *Caesar's Column* covers most of the hot political and social topics of the day. Hard to find but worth the search. (The two quotes above are from this book.)

Charles A. Lindbergh *Autobiography of Values.* Lindbergh did much to change the face of the world. He was the son of a prominent populist-Republican Minnesota congressman. Near the end, Lindbergh sought to find some meaning from his life. Did aviation serve humanity or wound it? Was the machine age worth the damage to the environment? In *Autobiography of Values,* Lindbergh ranges over a thousand interesting subjects with a wisdom that only comes with age. Truly one of populism's giants.

There are two modern books that could be called Neo-Populist: Kevin Phillips' *Politics of Rich and Poor* (the Democratic Party's new political handbook for the 1990s) and William Greider's *Secrets of the Temple* (the best and most comprehensive treatment of the Federal Reserve System and monetary policy yet written). Neither author is definitively Populist in the nineteenth-century sense, but then, nineteenth century Populism needs serious updating. Both books could be considered an excellent start on that process.